# Daily Warm-Ups

# GENERAL SCIENCE

Glen Phelan

## Level I

1 2 3 4 5 6 7 8 9 10
ISBN 0-8251-5946-6

J. Weston Walch, Publisher
P.O. Box 658 • Portland, Maine 04104-0658
walch.com
Printed in the United States of America

iii

**The *Daily Warm-Ups* series** is a wonderful way to turn extra classroom minutes into valuable learning time. The 180 quick activities—one for each day of the school year—practice science skills. These daily activities may be used at the very beginning of class to get students into learning mode, near the end of class to make good educational use of that transitional time, in the middle of class to shift gears between lessons—or whenever else you have minutes that now go unused.

*Daily Warm-Ups* are easy-to-use reproducibles—simply photocopy the day's activity and distribute it. Or make a transparency of the activity and project it on the board. You may want to use the activities for extra-credit points or as a check on the science skills that are built and acquired over time.

However you choose to use them, *Daily Warm-Ups* are a convenient and useful supplement to your regular lesson plans. Make every minute of your class time count!

## Name That Science

Science is the study of the natural world. That includes a lot of topics. The topics are grouped into different branches. You will study some of the topics in each branch of science.

Use the clues below to help you unscramble the words about different branches of science. Write the unscrambled word in the space provided. Then circle the number of the science in which you would learn about animals.

1. RATEH NSEEICC ________________ the study of the land, oceans, and weather
2. SYAPILCH NSEEICC ________________ the study of matter and energy
3. FLEI NSEEICC ________________ the study of living things
4. CAPSE NSEEICC ________________ the study of the sun, moon, planets, and stars
5. OOLEHCTYGN ________________ the use and application of scientific discoveries

# Step by Step

Scientists usually follow the scientific method to solve problems. The scientific method consists of five steps: identify the problem, gather information, make a hypothesis, test the hypothesis, and draw conclusions.

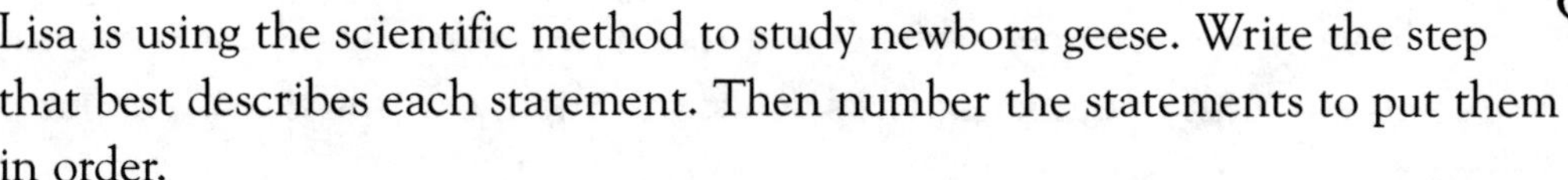

Lisa is using the scientific method to study newborn geese. Write the step that best describes each statement. Then number the statements to put them in order.

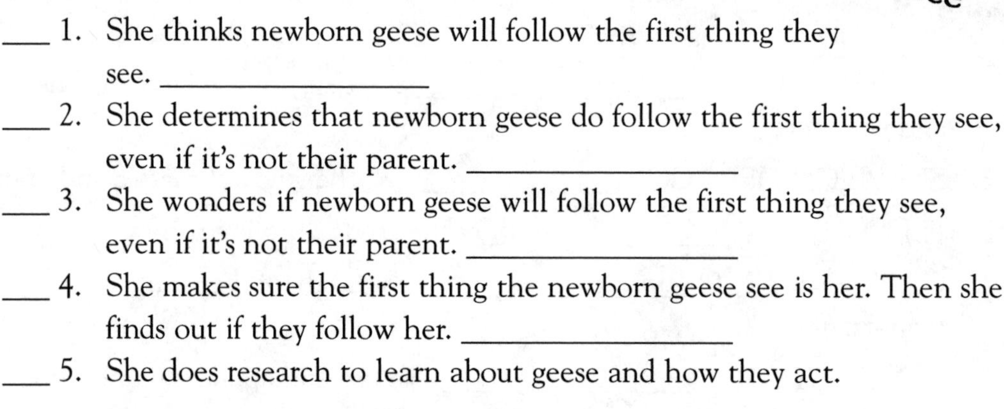

___ 1. She thinks newborn geese will follow the first thing they see. ____________________

___ 2. She determines that newborn geese do follow the first thing they see, even if it's not their parent. ____________________

___ 3. She wonders if newborn geese will follow the first thing they see, even if it's not their parent. ____________________

___ 4. She makes sure the first thing the newborn geese see is her. Then she finds out if they follow her. ____________________

___ 5. She does research to learn about geese and how they act.

____________________

Daily Warm-Ups: General Science

# Take a Try at SI

Scientists make measurements using a version of the metric system called the *International System*, or SI. Fill in the tables of SI measurements.

**SI Units**

| Unit | Symbol | Measures |
|---|---|---|
| gram | | mass |
| | m | |
| cubic meter | | |

**SI Prefixes**

| Prefix | Symbol | Means |
|---|---|---|
| milli- | m | |
| centi- | | 1/100 |
| | k | |

# The Right Tool

You probably have heard the expression "the right tool for the right job." This is certainly true when studying science. In science, you use many different tools to observe, measure, and explore.

Choose the tool from the box that should be used for each task described below. Write the letter of the tool on the line provided. You will NOT use all of the words.

| | | |
|---|---|---|
| a. balance | d. microscope | g. telescope |
| b. Bunsen burner | e. safety goggles | h. thermometer |
| c. graduated cylinder | f. stopwatch | |

___ 1. used to measure the volume of a liquid

___ 2. used to observe objects that are very small

___ 3. used to heat liquids

___ 4. used to measure the mass of an object

___ 5. used to protect your eyes from chemicals and sharp objects

# Choose Your Unit

Every measurement includes a number and a unit. The unit you choose must make sense with what you are measuring.

Choose the unit from the box that you would use for each measurement listed below. Write the letter of the unit on the line provided. Use each unit only once.

| | | |
|---|---|---|
| a. centimeter | d. kilogram | g. milliliter |
| b. degrees Celsius | e. kilometers per hour | h. millimeter |
| c. gram | f. liter | |

___ 1. mass of a small rock

___ 2. length of a leaf

___ 3. mass of a person

___ 4. temperature of water

___ 5. volume of liquid in a graduated cylinder

___ 6. width of a pine needle

___ 7. volume of an aquarium

___ 8. speed of a cheetah

## Safety First

Doing experiments is a fun and important part of learning science. But you must take special care when working with laboratory equipment.

Below are some important safety rules to follow when working in the science laboratory. Complete each rule by unscrambling the word or words in parentheses and writing them in the spaces provided.

1. Wear (ETSFAY LEGSGOG) ______________ whenever you are told to do so, such as when (ETGINAH) ______________ a liquid or using (RHSPA) ______________ objects.
2. Never used chipped (SWAGLAESR) ______________.
3. Wipe up all (LILPSS) ________________ immediately.
4. Tell your teacher right away if a(n) (DCICATNE) ______________ occurs.
5. Never mix (CCLAMSHEI) ______________ unless told to do so.
6. Never leave a(n) (SUNBNE NUBRRE) ______________ or a(n) (TOH ETALP) ______________ unattended.

## Getting to Know Metric

Most of the measurements you make in science will be in SI, a version of the metric system. You will use units such as centimeters, grams, and milliliters instead of inches, pounds, and gallons. To get used to metric, it helps to compare metric measurements to things you are familiar with.

Match each item in the second column with its closest measurement in the first column. Write the letter on the line provided.

| **Measurement** | **Item** |
|---|---|
| ___ 1. 1 kilometer | a. the width of your fingernail |
| ___ 2. 1 kilogram | b. the thickness of a dime |
| ___ 3. 1 centimeter | c. the length of five city blocks |
| ___ 4. 20 degrees Celsius | d. a hardcover dictionary |
| ___ 5. 1 millimeter | e. room temperature |
| ___ 6. 1 liter | f. three cans of soft drink |

# A Little Math

You can convert between metric and English measurements by using a conversion chart. Use the chart that follows to complete the conversions below.

1. 6 feet = ____________ meters
2. 8 liters = ____________ quarts
3. ____________ kilograms = 2.5 pounds
4. ____________ yards = 23 meters
5. 10 miles = ____________ kilometers
6. What is more, a gallon of milk or 4 liters of milk? ____________

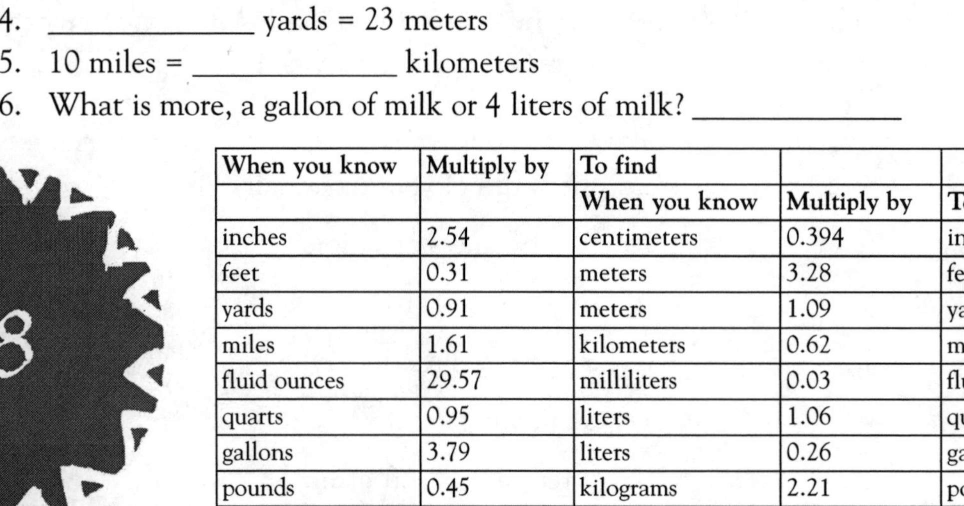

| When you know | Multiply by | To find | | |
|---|---|---|---|---|
| | | When you know | Multiply by | To find |
| inches | 2.54 | centimeters | 0.394 | inches |
| feet | 0.31 | meters | 3.28 | feet |
| yards | 0.91 | meters | 1.09 | yards |
| miles | 1.61 | kilometers | 0.62 | miles |
| fluid ounces | 29.57 | milliliters | 0.03 | fluid ounces |
| quarts | 0.95 | liters | 1.06 | quarts |
| gallons | 3.79 | liters | 0.26 | gallons |
| pounds | 0.45 | kilograms | 2.21 | pounds |

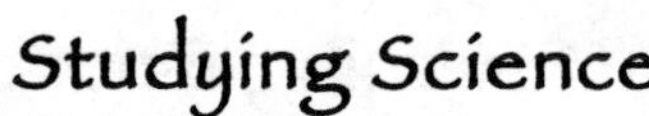


# What's My Career?

Have you ever thought that you might like to be a scientist? There are many different kinds of scientists. There are a lot of other science-related careers, too. And they do not all require advanced college degrees.

Read each description of a career. Then find and trace the name of the career in the box. Do not move diagonally or skip any squares.

1. I study conditions in the air to predict the weather.

| | | | | |
|---|---|---|---|---|
| T | E | O | R | O |
| E | L | G | D | L |
| M | O | M | Y | O |
| A | I | E | L | G |
| N | J | T | S | I |

2. I know all about the animals that live in the national park where I work.

| | | | | |
|---|---|---|---|---|
| D | P | B | L | C |
| T | S | H | N | K |
| E | P | A | R | K |
| V | A | X | Y | R |
| R | E | G | N | A |

3. I have to know a lot about the laws of motion in order to design planes.

| | | | | |
|---|---|---|---|---|
| R | B | A | E | R |
| T | U | A | N | O |
| I | C | A | L | E |
| E | N | I | G | N |
| E | R | W | I | S |

4. I work in a hospital where I take pictures of people's bones.

| | | | | |
|---|---|---|---|---|
| N | I | C | I | A |
| H | T | N | S | N |
| C | B | A | J | O |
| E | N | F | E | R |
| T | Y | A | R | X |

# What's Your Conclusion?

Andrew and Alma did an experiment to find out how different amounts of water affect the growth of a geranium plant. They planted four identical geranium plants. None of the plants had flowers. They cared for them the same except they watered them each a different number of times per week. After eight weeks, they had the data in the table below.

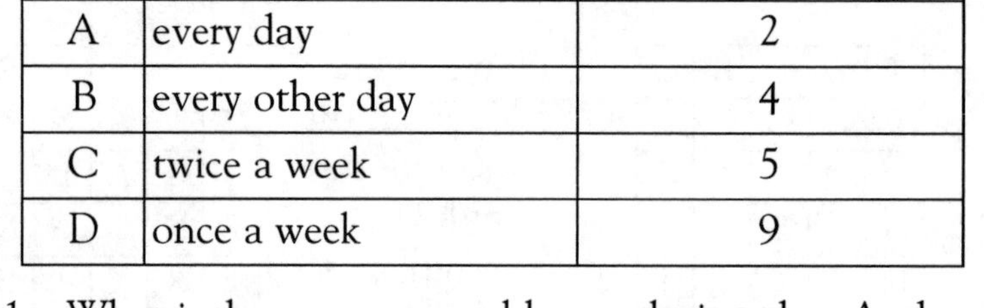

| Plant | Frequency of Watering | Number of Flowers |
|---|---|---|
| A | every day | 2 |
| B | every other day | 4 |
| C | twice a week | 5 |
| D | once a week | 9 |

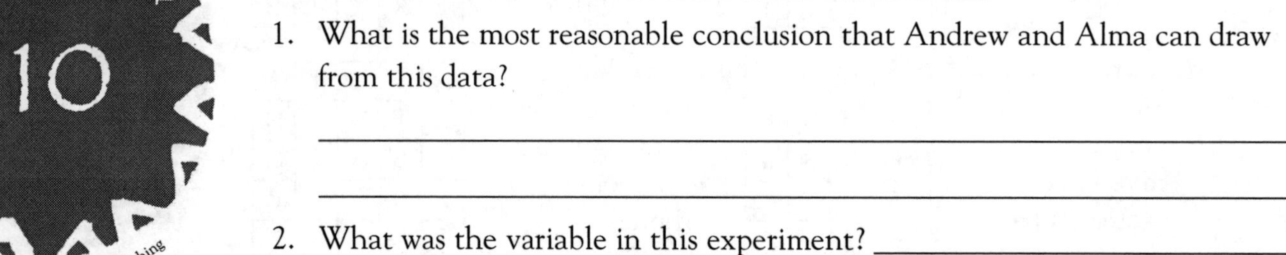



1. What is the most reasonable conclusion that Andrew and Alma can draw from this data?

   ______________________________________________

   ______________________________________________

2. What was the variable in this experiment? ____________________

# That's Life

How do you know if something is alive? Every living thing carries out certain activities called life processes. A living thing must carry out these processes to stay alive.

Match the life process in the second column with the correct example or description in the first column. Write the letter on the line provided. Then unscramble the underlined letters in the second column to find another word for a living thing.

**Example/Description**

___ 1. prevents a living thing from poisoning itself

___ 2. infant → toddler → adolescent → adult

___ 3. a lion eating a gazelle

___ 4. making more of the same kind

___ 5. leaves of a houseplant turning toward the sunlight

**Life Process**

a. getting and using energy

b. growing and developing

c. responding to changes

d. getting rid of wastes

e. reproducing

Another word for a living thing is ___ ___ ___ ___ ___ ___ ___ m .

Daily Warm-Ups: General Science

# Tell a Cell by Its Parts

The cell is the basic unit of all living things. You are made of billions of cells. So is a tree. Animal cells and plant cells have some similar structures. But there are some differences, too.

The drawing shows the cells of a plant and an animal. Identify the lettered structures of each cell. Use the words from the box and write them on the lines provided. You will not use all of the words.

| | | |
|---|---|---|
| cell wall | cytoplasm | ribosomes |
| cell membrane | mitochondria | vacuole |
| chloroplast | nucleus | |

a. ____________________

b. ____________________

c. ____________________

d. ____________________

e. ____________________

f. ____________________

Which cell is a plant cell? ____

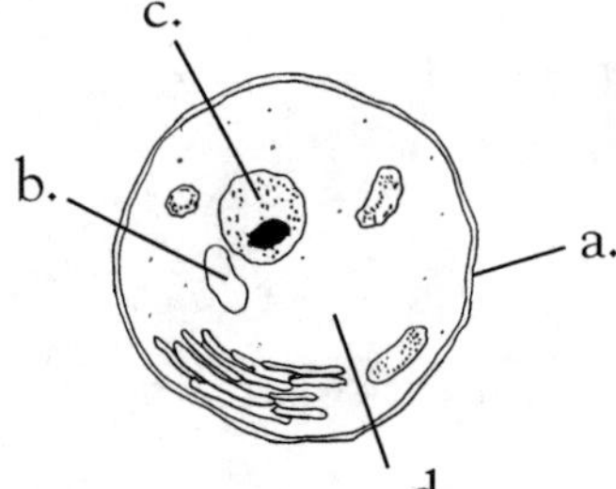

Cell 1

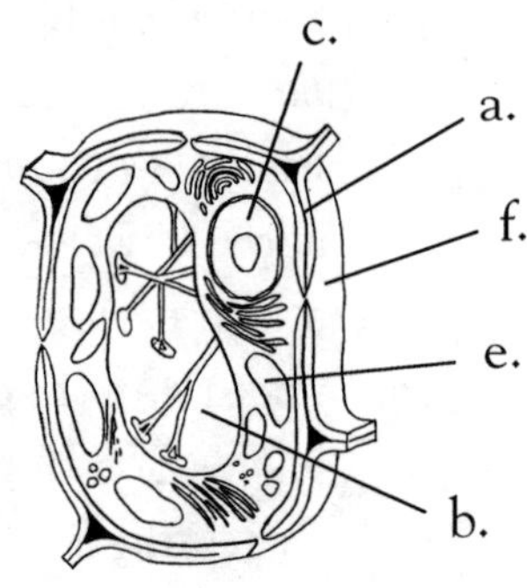

Cell 2

Daily Warm-Ups: General Science

# Find the Hidden Parts

If you look at a cell through a microscope, you will see that it is packed with structures. Each structure does a certain job that helps keep the organism alive.

You do not need a microscope to find the cell parts in the puzzle below. But you will still have to look carefully. Read each job of the cell part. Then circle the name of that part in the string of letters that follows. The name may read forwards or backwards.

1. Proteins are assembled here. S T A R I B O S O M E S Y L Y O P
2. They release energy from stored food. Y M I N M I T O C H O N D R I A M
3. It allows material to move in and out of the nucleus. E N A R B M E M R A E L C U N B I
4. They store water, salts, and other materials. R H O B I S A P E V A C U O L E S

Daily Warm-Ups: General Science

## Move It Out

Materials constantly move in and out of a cell. Food molecules move into the cell. Waste molecules move out.

The sentences describe how molecules move into and out of cells. Read each description. Then write the process by using the code. How? Match each number under the line to the pair of letters for that number. Then decide which letter to use.

| AB | CD | EF | GH | IJ | KL | MN | OP | QR | ST | UV |
|---|---|---|---|---|---|---|---|---|---|---|
| 1 | 2 | 3 | 4 | 5 | 6 | 7 | 8 | 9 | 10 | 11 |

| | |
|---|---|
| 1. the movement of molecules from an area of higher concentration to an area of lower concentration | ___ ___ ___ ___ ___ ___ ___ ___ ___<br>2 5 3 3 11 10 5 8 7 |
| 2. the movement of water through a cell membrane | ___ ___ ___ ___ ___ ___ ___<br>8 10 7 8 10 5 10 |
| 3. the movement of molecules from an area of lower concentration to an area of higher concentration (two words) | ___ ___ ___ ___ ___ ___<br>1 2 10 5 11 3<br>___ ___ ___ ___ ___ ___ ___ ___ ___<br>10 9 1 7 10 8 8 9 10 |

## Be a Square

An organism receives, or inherits, a gene for a trait from each parent. A pea plant might inherit a gene for round seeds (R) from one parent plant and a gene for wrinkled seeds (r) from the other parent plant. Then the offspring plant will produce round seeds. If the offspring inherits two genes for wrinkled seeds (rr), it will produce wrinkled seeds.

A Punnett square shows the possible combinations of genes for a trait. Complete the Punnett square. Fill in the gene combinations for the four offspring. How many of the offspring will produce round seeds?

__________

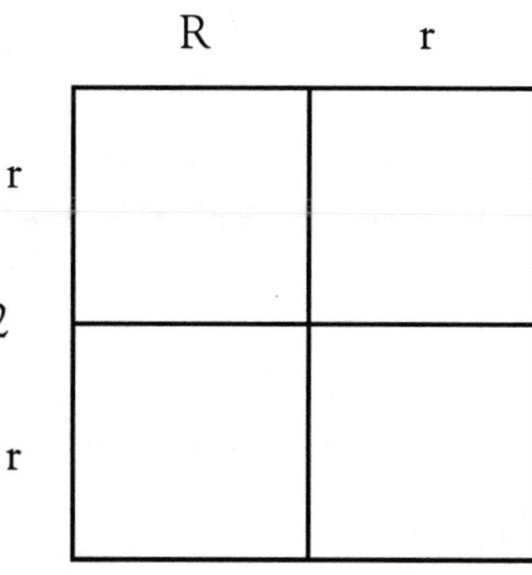

## What Are the Chances?

A Punnett square is a useful tool for predicting the traits of offspring. However, you cannot always predict with certainty. A Punnett square shows you only what the *chances are* of an offspring having certain traits. In the Punnett square below, neither parent has sickle cell anemia, but they each have a gene for sickle cell anemia. They are carriers for the disease. A child must have the sickle cell gene from each parent to have the disease. Complete the Punnett square. What are the chances that a child will have sickle cell anemia? What are the chances a child will not? Your answers should be written as percentages, such as 25%.

S = no sickle cell anemia

s = sickle cell anemia

Chance of no disease: ________

Chance of disease: ________

Parent 1 (columns), Parent 2 (rows)

| | S | s |
|---|---|---|
| **S** | | |
| **s** | | |

## One Step at a Time

An organism grows when its cells divide. Before a cell divides into two cells, it makes copies of its chromosomes. Then the cell goes through a process called *mitosis*.

The diagram shows a cell that has just made copies of all its chromosomes. It is now ready to go through mitosis. In each circle, diagram each step of the process. Use the captions under the circles as clues.

chromosomes copied → chromosomes line up → chromosomes separate → new cells form

In the last step of mitosis, something forms around each new group of chromosomes. Unscramble the letters to find out what it is.

CURAENL ENMMBAER

________________ ________________

Daily Warm-Ups: General Science

## More Steps

The process of meiosis takes place in two stages. The steps in each stage are listed below, but they are out of order. Put the steps in the correct order by writing the numbers 1–5 in the spaces before the steps for the first stage. Write the numbers 6–10 for the second stage. Assume the parent cell has already made copies of its chromosomes.

**First Stage**

___ Nuclear membrane forms around each new group of chromosomes.

___ Pairs of twin chromosomes line up.

___ Cell divides in two.

___ Pairs of twin chromosomes move apart to opposite sides of cell.

___ Nuclear membrane disappears.

**Second Stage**

___ Twin chromosomes move apart to opposite sides of cell.

___ Twin chromosomes line up.

___ The cell divides in two.

___ Nuclear membrane in each cell disappears.

___ Nuclear membrane forms around each new group of chromosomes.

# The Complete DNA

In 1953, James Watson and Francis Crick made one of the most important scientific discoveries of all time. They discovered the structure of DNA.

The paragraph below about DNA is incomplete. Two of the four numbered sentences complete the paragraph best. Choose the two sentences and circle them.



A DNA molecule is shaped like a twisted ladder. Each rung of the ladder is made of a pair of chemicals called bases. When DNA copies itself, it splits down the middle of its rungs. The base pairs separate.

1. When DNA makes a copy of itself, it is called replicating.
2. Then new bases pair with the bases on each half of the DNA molecule.
3. DNA contains all the information that gives an organism its traits.
4. When all the bases finish pairing up, two identical DNA molecules have been formed.

# A Genetics Crossword

Like any area of science, the study of genetics has its own set of vocabulary words. Use the definitions to fill in this crossword puzzle with the correct genetics words.

**Across**

3. a characteristic of an organism
4. the set of genes for a given trait in an organism
5. a gene that is expressed in an organism
6. a gene that is not expressed in an organism

**Down**

1. the appearance of an organism
2. threadlike strands containing DNA in a cell's nucleus
4. Thousands of these make up each human chromosome.

# Lots of Changes

Millions of different kinds of organisms exist. This diversity is the result of a process that takes place over time. Use the clues to complete the puzzle and discover the name of this process.

1. theory that says more recent species are changed descendents of earlier species: ________________ with modification
2. A species does this as it changes over time.
3. an inherited trait that helps a species survive
4. Scientists often study these to learn how species have changed over time.
5. theory that says organisms best suited to the environment survive to pass on their genes: ________________ selection
6. a change in the DNA that makes up a gene
7. a group of organisms that naturally interbreed and produce offspring that can reproduce
8. different form of a trait in a species
9. developed theories about how species change over time

1. _ _ _ _ [_] _ _
2. _ _ _ _ [_] _
3. _ _ _ _ _ _ _ _ [_] _
4. _ _ _ _ _ [_] _
5. _ _ _ [_] _ _ _
6. _ _ _ _ [_] _ _ _
7. _ _ _ _ [_] _ _
8. _ _ _ _ _ _ _ [_] _
9. _ _ _ _ _ [_]

# Are We Related?

Scientists compare body parts of different organisms to determine how they are related. *Homologous structures* are body parts that are in the same position in different species. *Vestigial structures* are body parts that seem to be useless but probably were useful to the organism's ancestors. *Analogous structures* have a similar function in different species but are in different positions.

Write H, V, or A to show whether each structure or structures are homologous, vestigial, or analogous.

1. ____ wings of bird and wings of insect
2. ____ bones of human arm and bones of whale flipper
3. ____ human tailbone
4. ____ bones of bat wing and bones of bird wing
5. ____ eyes in blind cave crayfish

# What's the Evidence?

If species change over time, there should be evidence of it. There is. Complete the table below to summarize some of the evidence for evolution.

| Kind of Evidence | Example | What Evidence Shows |
|---|---|---|
| fossil record | | |
| similar stages of development | Embryos of humans, chickens, and other vertebrates are similar. | |
| | bones of human arm and bones of cat leg | Four-limbed animals with backbones shared a common ancestor. |

Daily Warm-Ups: General Science

# Four Points to Remember

In the mid-1800s, Charles Darwin wrote his book *On the Origin of Species*. In it, he explained his theory of evolution by natural selection. Since that time, much evidence has supported his theory.

Darwin's theory can be summarized as four main points. Match each point of the theory with the correct example. Write the letter in the space provided.

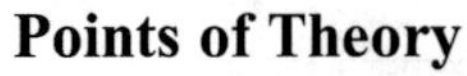

**Points of Theory**

1. ___ Organisms produce more offspring than can survive.
2. ___ Variation exists among individuals in a population.
3. ___ Individuals within a population struggle to survive.
4. ___ Individuals that are best suited to the environment are more likely to survive and pass their genes to their offspring.

**Examples**

a. Gazelles that can avoid being killed by lions have a chance to mate and reproduce.
b. A population of cardinals may differ slightly in color, size, and speed.
c. Fish lay thousands of eggs at a time.
d. The roots of corn plants compete for water and nutrients.

# A Changing System

When scientists first started classifying life, they thought all organisms were either plants or animals. Then they realized that microscopic organisms were different enough to be given their own kingdom called protists. Later, mushrooms, yeasts, and molds were separated from the plants and called fungi. Then scientists placed some microorganisms in their own kingdom. They were called monerans. Now there were five kingdoms. But recent discoveries have led scientists to break the monerans into eubacteria and archaebacteria.

Fill in the chart below to show how the number and names of the kingdoms have changed over the centuries. Start with the oldest classification system on top.

<table>
<tr><td colspan="5"></td><td></td></tr>
<tr><td colspan="3"></td><td colspan="2"></td><td></td></tr>
<tr><td colspan="2"></td><td></td><td></td><td></td><td></td></tr>
<tr><td></td><td></td><td></td><td></td><td></td><td></td></tr>
</table>

# Practice Classifying

Venn diagrams are helpful tools for classifying organisms. In fact, you can use these diagrams to classify anything.

The Venn diagram below shows how some books in a library could be classified. Areas that overlap have something in common. Use the following terms to label the diagram: all books, cookbooks, Japan guidebooks, mysteries, travel books.

A. ________________

B. ________________

C. ________________

D. ________________

E. ________________

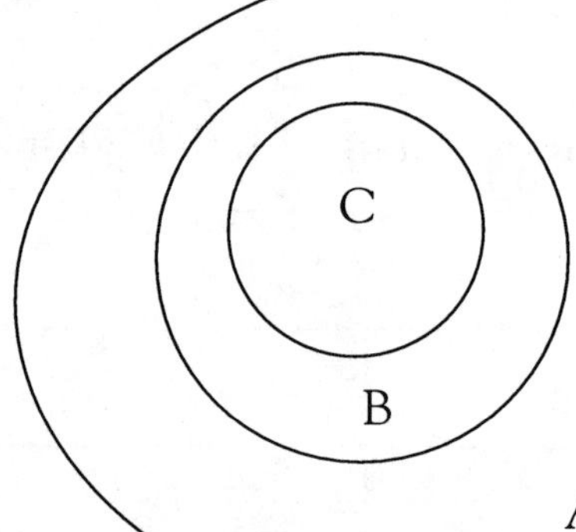

E

A

# Everything in Its Place

Biologists use a seven-level system to classify organisms. Each level, from largest to smallest, includes fewer and fewer organisms until you reach the last level, which includes only one kind of organism.

Fill in the classification pyramid with the names of the levels. One is done for you. Then make up a sentence whose words use the first letter of each level to help you remember the levels in order.

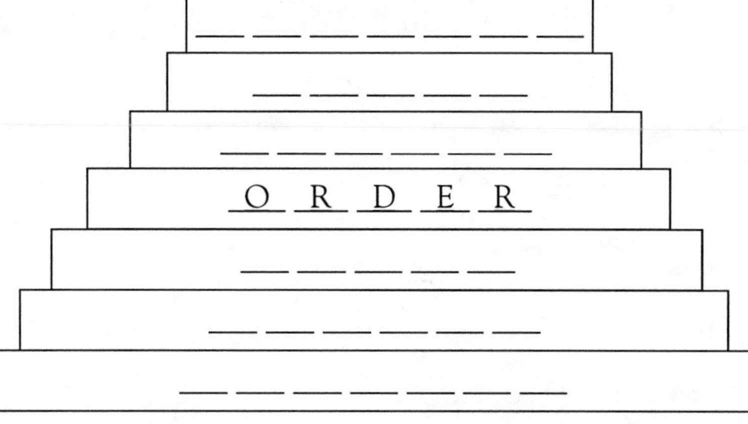

# Classify Your Address

In the Linnaean classification system, each level includes fewer and fewer organisms. The kingdom level includes more organisms than the phylum level, the phylum level includes more than the class level, and so on.

Classifying an organism is a lot like classifying an address. You might start with the continent, then go to the country, state, county, city, street, and specific address. Fill in the lines below to classify your address.

continent ____________________________

country ____________________________

state ____________________________

county ____________________________

city ____________________________

street ____________________________

address ____________________________

# The Shopping List

Classifying a new organism can be a difficult task. But you already have lots of practice at it. When you shop for a CD or a pair of shoes, you classify. When you look for a birthday card, you classify.

Make a classification system for the grocery shopping list below. Decide what categories to use for the various items. Then list the appropriate items in those categories. *Hint:* Think about where you would go in the store to find the items.

**Shopping List**

| | | |
|---|---|---|
| yogurt | apples | cheddar cheese |
| chicken legs | lettuce | fresh string beans |
| frozen corn | hamburger | pizza |
| ice cream | milk | watermelon |

# The Same, Yet Different—Bacteria

When scientists invented the microscope, it opened an entire new world. For the first time, people saw that microscopic life is all around us. The most common microorganisms are bacteria.

Compare and contrast concepts of bacteria. Tell how the following pairs of terms are alike and different.

1. virus : bacterium

   alike: ______________________________

   different: ______________________________

2. eubacteria : archaebacteria

   alike: ______________________________

   different: ______________________________

3. bacillus : spirillum

   alike: ______________________________

   different: ______________________________

# Up, Up, and Away

Many bacteria reproduce by replicating their DNA and then dividing in half. Each bacterium produces two identical bacteria. If conditions are favorable, a population of bacteria can grow quickly. The table shows that the number of bacteria doubles every half-hour. Complete the table. Then use the table to make a graph showing how the population grew.

| Hours | 0 | $\frac{1}{2}$ | 1 | $1\frac{1}{2}$ | 2 | $2\frac{1}{2}$ | 3 | $3\frac{1}{2}$ | 4 | $4\frac{1}{2}$ | 5 |
|---|---|---|---|---|---|---|---|---|---|---|---|
| Number of bacteria | 1 | 2 | 4 | | | | | | | | |

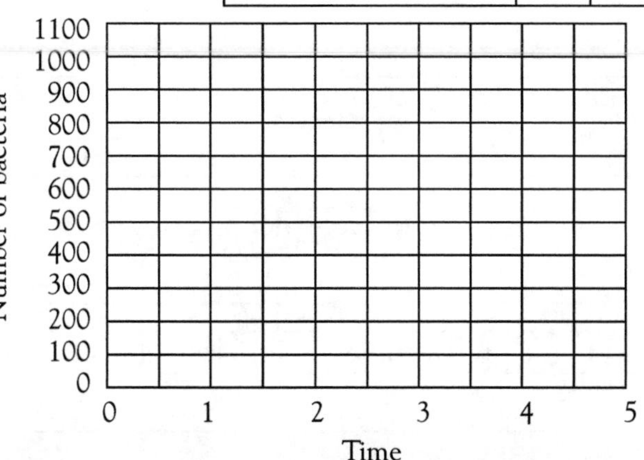

## Not All Bad

When you hear the word *bacteria*, you might think of only harmful things. Some bacteria do cause disease. However, most bacteria are either harmless or helpful to people.

Arrange each group of words to form a sentence about how bacteria are helpful. Write the complete sentence on the line provided. Be sure to punctuate your sentences correctly.



1. plant roots | bacteria in | into compounds | change nitrogen | can use | the plants

   ______________________________

2. produce | cheese, yogurt, and sour cream | are used to | bacteria | foods such as

   ______________________________

3. help | organisms | bacteria | soil by | dead | produce | decomposing

   ______________________________

4. oil | up oil | and are used to | digest | spills | some bacteria | clean

   ______________________________

# Protist Junk Drawer

Do you have a junk drawer at home? It probably has a variety of objects in it. Maybe the only thing the objects have in common is that they are small enough to fit in a drawer. The protist kingdom might be called the "junk-drawer-kingdom." Protists are diverse. The only things they have in common with one another is that they all have cells with nuclei and live in wet environments.

Match each group of protists in column A with their description in column B. Write the correct letter in the space provided.

**Column A**

___ 1. protozoans

___ 2. downy mildew

___ 3. green algae

___ 4. brown algae

___ 5. diatoms

**Column B**

a. funguslike; unicellular; often attack food crops

b. plantlike; includes underwater "forests" of giant kelp

c. animallike; unicellular; can move around

d. plantlike; unicellular; used in toothpastes and household scouring products

e. plantlike; have green pigments; includes some types of seaweed

## Fungus Among Us

The fungus kingdom includes more than 100,000 species. They live in water, soil, and even inside other organisms. Fungi cannot make their own food. They get food by breaking down waste or dead matter.

Use the clues to fill in the missing words about fungi. Then match the numbers beneath the letters to find the answer to the question below.

1. Fungi grow by producing these fine threads. __ __ __ __ __ __ (1 under the 3rd blank, 2 under the 6th blank)

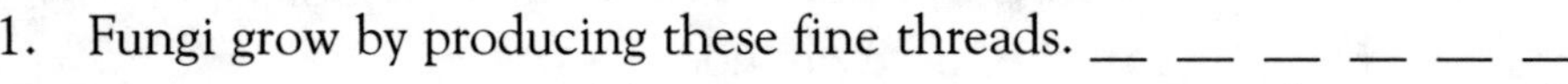

2. Club fungi that people often eat. __ __ __ __ __ __ __ __ __ (3 under the 1st blank, 4 under the 2nd blank)

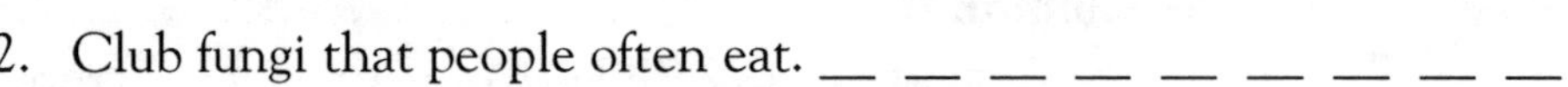

3. Fungus that often grows on old bread. __ __ __ __ (5 under the 3rd blank)

4. A mass of hyphae. __ __ __ __ __ __ __ __ (6 under the 3rd blank, 7 under the 5th blank, 8 under the 6th blank)

This fungus was used to make the first medicine to kill harmful bacteria.

__ __ n i __ i __ __ __ __ __

1 2 _ _ 6 _ 5 7 8 4 3

# Plant Words

What words come to mind when you think of plants? Maybe you think of some of the words listed here. Find and circle the words from the box in the puzzle below. You may find the answers by reading up, down, forward, backward, or diagonally. Then use some of the remaining letters to write an important plant process in the spaces provided.

| | | | |
|---|---|---|---|
| fruit | stem | seed | pollen |
| pistil | germination | sunlight | stamen |
| root | plant | leaf | water |

M P P I S T I L P F R U I T S
E H O T O S Y N E O T P H E T
T S I S K B N Q S A L E S E A
S U N L I G H T S A F L E N M
B B S N O I T A N I M R E G E
U W A T E R T T O O R N D N N

Plant process: ___ ___ ___ ___ ___ ___ ___ ___ ___ ___ ___ ___ ___ ___

## Flower Sudoku

In a sudoku puzzle, each three-by-three box and each row and column in a nine-by-nine grid has to have the numbers 1–9. You can play sudoku with words, too. Solve the puzzle for the word *flowering*. The middle box is done for you.

| | | | | | | | | |
|---|---|---|---|---|---|---|---|---|
| | | | N | | | | I | G |
| | N | | | W | I | | | |
| W | I | | | L | R | O | N | |
| | | N | L | G | F | | R | |
| G | R | | I | O | W | | E | L |
| | W | | R | N | E | G | | F |
| | L | O | E | I | | | W | N |
| | | | W | R | | | F | |
| R | E | | | | N | | | |

## Plant Math

Carbon dioxide enters a leaf, and oxygen and water exit a leaf, through tiny pores. What are these pores called? To find the name, use some math.

First, divide 2 into each of the numbers in Table A. Write the answer above the dividend. Then match the answer to the letter in Table B. The first one is done for you.

**Table A**

| | | | | | | | |
|---|---|---|---|---|---|---|---|
| | S | | | | | | |
| | 3 | | | | | | |
| 2 | 6 | 8 | 4 | 0 | 2 | 8 | 2 |

**Table B**

| | |
|---|---|
| 0 | M |
| 1 | A |
| 2 | O |
| 3 | S |
| 4 | T |

Daily Warm-Ups: General Science

## It's a Process

Think of a beautiful flower. The flower is the result of several processes the flowering plant goes through. Fill in the spaces below to name some of these processes. Use the clues to help you.

1. __ __ __ __ __ __ation (early growth)
2. __ __ __ __ __ __ation (insects and birds help)
3. __ __ __ __ __ __ __ __ation (beginning of a seed)

What is the correct order of the three processes above? ______________________

# In the Right Place

What would happen if you removed a cactus from a desert in Arizona and planted it in a forest in Georgia? Chances are, the cactus would not do very well. It is adapted to desert conditions.

A cactus has a thick stem, spines, and deep roots. Think about these features. Then think about desert conditions: hot, little rainfall, sandy soil through which water sinks quickly. Explain how the cactus is adapted to the desert by answering the questions.

1. Why does a cactus have a thick stem?

______________________________________________

2. How do deep roots help the cactus?

______________________________________________

3. Spines are actually leaves. But they don't carry on transpiration. Why?

______________________________________________

## Dial Up a Response

What happens if you set a houseplant in a sunny window? Over several days, its leaves, stems, and flowers will grow toward the light. This is one way plants respond to their environment. Some plants respond to touch. A vine might coil around anything it touches.

There is a name for these kinds of plant growth responses. Use the phone number below to "dial up" the name. Choose from among the letters on a telephone that match each number.

___ ___ ___ ___ ___ ___ ___

8 7 6 – 7 4 7 6

# Tasty Plants

Did you know that a white potato is a stem? You probably eat lots of different plant parts. Try classifying the foods on the grocery list according to their plant parts.

**Grocery List**

| | | | |
|---|---|---|---|
| asparagus | cauliflower | kidney beans | radishes |
| broccoli | celery | lettuce | spinach |
| cabbage | cherries | oranges | sweet potatoes |
| carrots | peanuts | corn | tomatoes |

| stems | leaves | roots | seeds | flowers | fruits |
|---|---|---|---|---|---|
| | | | | | |

Daily Warm-Ups: General Science

# What Arthropod Am I?

What animal phylum is the most successful of all time? Arthropods! At least 750,000 species of arthropods have been identified. That's more than three times all other animal species combined! All arthropods have an outer skeleton, jointed legs, and a segmented body.

Some groups of arthropods are listed in the box below. Choose from among these groups to answer the riddles.

| a. crustacean | c. centipede | e. insect |
|---|---|---|
| b. arachnid | d. millipede | |

**Riddle 1**

I have three body sections, six legs, and one pair of antennae. I have wings, but not all of us do. What am I? _______

**Riddle 2**

I have two or three body sections and two pairs of antennae. Most of us live in water. Some of us make a fine meal at a seafood restaurant. What am I? _______

**Riddle 3**

I have two body sections. I have no antennae but eight legs. What am I? _______

## Lots of Changes

As insects grow and develop, they undergo a process in which they change their shape and form. Solve the word equation to figure out the name of this process.

(meet – e) + a + (more – e) + (phono – no) + s + is =

__ __ __ __ __ __ __ __ __ __ __ __ __

What are the two different types of this process?

________________________________________

________________________________________

# Put Some Backbone Into It, or Not

Animals with a backbone are called *vertebrates*. Animals without a backbone are *invertebrates*. A backbone is part of an internal skeleton that supports and protects the body.

Find three vertebrates and three invertebrates hidden in the word search puzzle. Then write them under the correct heading.

| | | | | | | | | |
|---|---|---|---|---|---|---|---|---|
| C | E | L | O | B | S | T | E | R |
| Y | L | F | R | E | T | T | U | B |
| L | A | G | F | Y | M | R | O | W |
| A | H | H | U | M | A | N | N | G |
| M | W | O | L | F | X | O | L | P |

**Invertebrates** ______________________ ______________________ ______________________

**Vertebrates** ______________________ ______________________ ______________________

## Which Doesn't Belong?

There are seven classes of vertebrates: jawless fishes, cartilaginous fishes, bony fishes, amphibians, reptiles, birds, and mammals.

Each set of examples below includes three animals from one class and one animal from a different class. Circle the animal that doesn't belong. Then write the name of the class to which the other three belong.

**Class**

1. turtle frog snake alligator ______________________
2. salamander frog newt whale ______________________
3. human horse bat penguin ______________________
4. blue jay eagle rabbit penguin ______________________
5. shark trout salmon swordfish ______________________

## Mammal Chart

Mammals are animals that are vertebrates, have hair or fur, and are fed milk produced in the mother's body. The 6,000 or so different species of mammals are classified into three groups, depending on how the young develop.

Fill in the chart with the number of the correct definition for each mammal group. Then fill in one example for each group.

| Mammal Group | Definition | Example |
|---|---|---|
| monotremes | | |
| marsupials | | |
| placentals | | |

**Definitions**

1. Young are born alive but must develop further in their mother's pouch.
2. Young develop fully inside their mother's body.
3. Mother lays eggs.

# A World of Adaptations

An *adaptation* is an inherited trait that lets a species survive in its environment. An adaptation might be structural, which involves body shape or color, or it might be behavioral, which involves how the animal acts.

Unscramble the letters to discover some of the many adaptations in the animal world.

1. A frog's color lets it blend into its surroundings. Its color is a form of this.
   OFCAAEMLUG ____________________
2. A Venus's flytrap is a plant that grows in soil that lacks certain nutrients. It gets the nutrients it needs by catching and "eating" these kinds of animals.
   CESNTIS ____________________
3. The "parachute" on a dandelion seed is an adaptation that lets the seeds be dispersed by this.
   IDWN ____________________

# Oh, Behave!

When you are told to behave, it means "be good." In science, when an organism behaves, it is just reacting to changes in its environment. So animal *behavior* is the way an animal reacts to changes in its environment.



Below is a pair of terms about behavior. Tell how the terms are alike and how they are different.

**innate learned**

alike: ______________________________________________

______________________________________________

different: ______________________________________________

______________________________________________

48

# A Map for Learning

Think about a skill that you have learned. Maybe it was riding a bike or hitting a baseball. Animals learn how to do things, too. Fill in the concept map to show three major kinds of learning and an example of each.

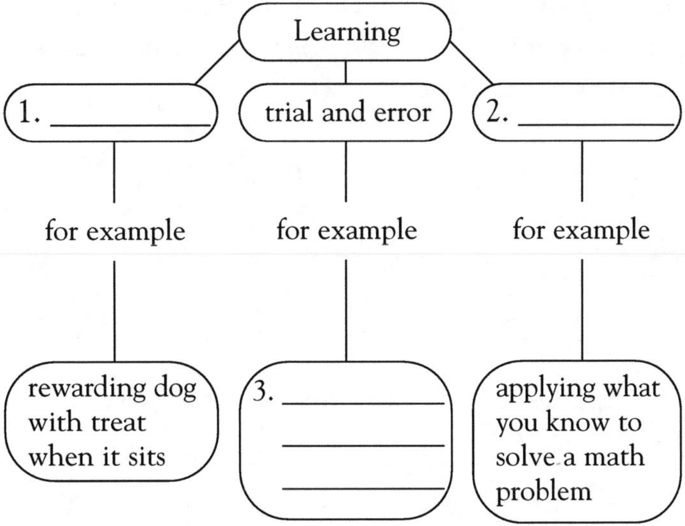

Learning

| 1. ____________ | trial and error | 2. ____________ |
|---|---|---|
| for example | for example | for example |
| rewarding dog with treat when it sits | 3. ____________ ____________ ____________ | applying what you know to solve a math problem |

## On the Brink

When the last member of a species dies, that species is *extinct*. It is gone forever. Governments have passed laws to protect species from extinction by human activities. As part of this protection, species that may be facing extinction are put into two categories. Each category is described below and hidden in the word search. The words may be read forward, backward, or around corners. Circle the words and write them on the correct lines below.

| | | | | | | |
|---|---|---|---|---|---|---|
| U | V | W | X | D | E | R |
| E | R | H | T | P | R | E |
| A | T | E | N | E | D | G |
| O | S | E | N | D | A | N |

1. Species that are declining in numbers so fast they may soon become extinct: ___________________
2. Species that are not declining in numbers as quickly but may be in trouble if something isn't done to increase their populations: ___________________

## Get Organized

The human body, like the bodies of other multicellular organisms, is made of millions of cells. But the cells are not just randomly spread throughout the body. They are organized.

Fill in the flow chart below to show the different levels of organization in the body. Use the words in the box. Then give an example of each level of organization.

| organ | system | tissue |
|---|---|---|

| Level of Organization | Example |
|---|---|
| cells<br>↓<br>1. ______________<br>↓<br>2. ______________<br>↓<br>3. ______________<br>↓<br>organism | 4. ______________________<br><br>5. ______________________<br><br>6. ______________________ |

## Name That System

Your body is made of many systems that work together to keep you healthy. Fill in the chart to list some of the body's main systems and some of the organs and jobs of each system.

| System | Main Organs | Main Jobs |
|---|---|---|
| skeletal | | supports and protects the body |
| | muscles | |
| digestive | | breaks down food |
| | | sends blood throughout the body |
| respiratory | lungs | |
| nervous | | |

## Mr. Bones

The human body has 206 bones that connect together to form a skeleton. The skeleton supports the body and protects organs, such as the brain and lungs. The skeleton also produces blood cells and provides places for muscles to attach.

Use the clues to identify words that have to do with the skeletal system. If you identify each word correctly, the circled letters will spell out the word *skeleton*.

1. bone that protects the brain: ◯_ _ _ _
2. made of many bones that protect your spinal cord: _ _ _ ◯_ _ _ _ _
3. tough bands of tissue that hold bones together: _ _ _ _ _ ◯_ _ _
4. mineral found in bones: _ _ ◯_ _ _ _ _
5. bones that make up the backbone: _ ◯_ _ _ _ _ _ _ _ _
6. tough band of tissue that connects muscle to bone: ◯_ _ _ _ _ _
7. place where two bones meet: _ ◯_ ◯_

# The Path of Food

From the moment you put a morsel of food in your mouth, your digestive system goes to work. This system breaks down the food into molecules that your cells can absorb. During digestion, the food follows a path that involves several organs and processes.

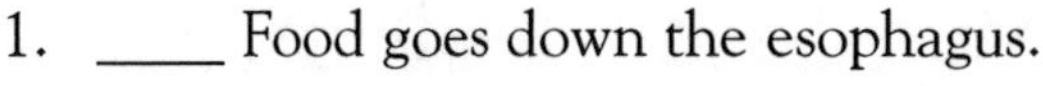

Below are some of the steps of digestion. Put the steps in the correct order by writing the numbers 1, 2, 3, and so forth in front of each step.

1. ____ Food goes down the esophagus.
2. ____ The gall bladder releases bile that breaks down fat droplets.
3. ____ Food enters the mouth and is chewed.
4. ____ The tongue moves food to the back of the throat.
5. ____ Undigested material passes through the large intestine.
6. ____ Saliva moistens food.
7. ____ The soupy food mixture enters the small intestines.
8. ____ Stomach juices break down food.

## Circulatory Password

Do you know how to play the password game? Usually, you are given one word at a time as a clue to figure out the mystery word. In this version of the game, you will be given an entire series of words. Read through each series, one word at a time, to figure out the mystery word. Each mystery word has something to do with the circulatory system.



1. chambers
   organ
   beats
   ventricles

**Mystery Word:**

__ __ __ __ __

2. plasma
   white cells
   platelets
   red cells

**Mystery Word:**

__ __ __ __ __

3. arteries
   veins
   capillaries
   flow

**Mystery Word:**

__ __ __ __ __ __ __

## Word Pairs

Your respiratory system moves oxygen into your body's cells and removes carbon dioxide from the cells. Your cells need oxygen in order to release energy. Your cells also need to get rid of carbon dioxide so it doesn't build up and become poisonous.

Each pair of terms below are important parts or processes involving the respiratory system. For each pair, write a sentence that uses the terms correctly.

1. inhale/exhale

______________________________________________

2. nose/mouth

______________________________________________

3. bronchi/lungs

______________________________________________

## You've Got Some Nerve

Your nervous system coordinates the movement and functions of all your body parts. This system is the body's communication network. The nervous system is made of two parts: the central nervous system and the peripheral nervous system.

In the puzzle below, circle the words that fill in the blanks of the sentence. The missing words run in a continuous line going up, across, or down.

The central nervous system consists of the _ _ _ _ _ and the _ _ _ _ _ _ _ _ _ _ _ _ _.

N S P I N
I D R O A
A R B C L

# The Eyes Have It



How do you know what's going on around you right now? You use your senses. Your senses include sight, hearing, touch, taste, and smell. Each sense depends on sense organs. Your eyes are the organs for your sense of sight.

How much do you know about the parts of the eye? Let's find out. Use the words in the box to label the cross section of the eye. Write the correct word on the line provided.

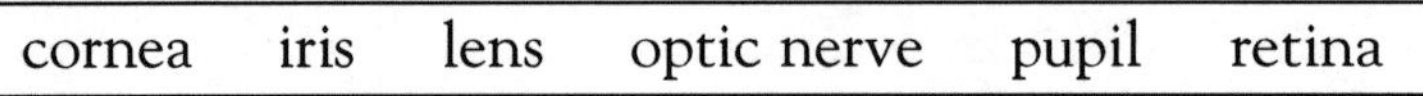

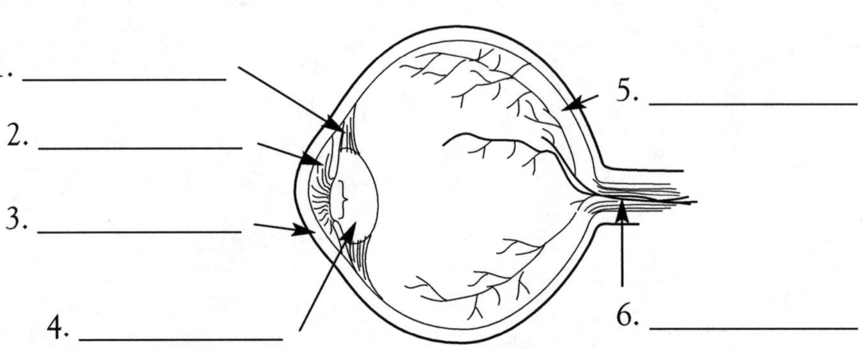

## Virus Attack

*Viruses* are tiny particles that cause disease. Colds, the flu, chicken pox, and AIDS are some of the diseases caused by viruses.

The steps below describe how viruses destroy cells and harm the body. However, the steps are out of order. Put them in the correct order by writing the numbers 1, 2, 3, and so forth in front of each step.

1. ____ The virus parts form new viruses.
2. ____ A virus injects genetic material into the cell.
3. ____ A virus attaches to the outside of the cell.
4. ____ The cell bursts and releases new viruses.
5. ____ The genetic material directs the cell to make new virus parts.

## Which One Doesn't Belong?

Let's review some of the body's systems. Each body system below is followed by the names of four organs or other body parts. Cross out the organ or part that doesn't belong in that system.

1. skeletal system: skull gallbladder joints vertebrae
2. digestive system: stomach saliva small intestine diaphragm
3. circulatory system: blood heart aorta liver

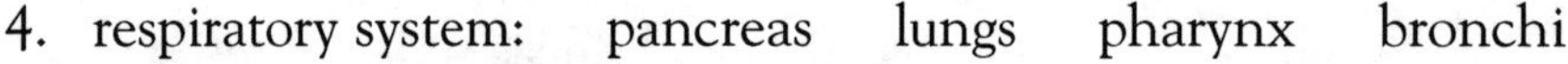

4. respiratory system: pancreas lungs pharynx bronchi
5. excretory system: kidneys ureter ventricles urinary bladder
6. nervous system: neuron spinal cord brain stem sperm
7. reproductive system: uterus hemoglobin testes egg

# You Are What You Eat

Why does your body need food? Because food provides the nutrients your body needs to get energy, repair itself, and carry out all the processes of life. There are thousands of different kinds of foods. But there are only six kinds of nutrients.

Circle the six kinds of nutrients hidden in the word puzzle below. The words can be found across, down, forward, or backward.

R C A R B O H Y D R A T E S
E Z Y S U V E N D E R B A X
T A P R O T E I N S I Q J Y
A B S S N I M A T I V O V F
W O U L P M J G C X A N M L
S T A F T M I N E R A L S Y

## Keys to Good Nutrition

Some keys to good nutrition are stated below. Complete each statement by using a word from the box. Write the letter of the correct word on the line provided.

| | | |
|---|---|---|
| a. Value | d. Pyramid | g. fiber |
| b. water | e. calories | h. variety |
| c. vegetables | f. label | i. calcium |

1. You should take in about 2 liters of ____ every day.
2. Drink milk to get the mineral ____ for strong bones.
3. Use the Food Guide ____ to help you choose foods.
4. It's not a nutrient, but your diet should include some foods with ____.
5. If you take in more ____ than you use, you will gain weight.
6. A healthy diet includes lots of fresh ____.
7. You can tell how nutritious a food is by reading the food ____.
8. Look at the Percent Daily ____ to see how much of a nutrient that food supplies in your daily diet.
9. The main key to good nutrition is to eat a ____ of foods.

# Disease Fighters

Beginning about 200 years ago, people started learning a lot about infectious diseases—what causes them and how to fight them. Match each disease fighter with his or her accomplishment. Write the letter of the correct disease fighter on the line.

____ 1. performed the first vaccination to protect against smallpox

____ 2. pushed to keep hospitals clean; founded the modern nursing profession

____ 3. developed techniques to prevent infections during surgery

____ 4. proved that microorganisms cause disease

____ 5. showed that a specific microorganism causes a specific disease

____ 6. discovered penicillin, which became the first antibiotic

a. Alexander Fleming

b. Edward Jenner

c. Robert Koch

d. Joseph Lister

e. Florence Nightingale

f. Louis Pasteur

# Build Some Healthy Habits

There are lots of things you can do every day to help prevent disease and stay as healthy as possible. Arrange the word blocks in each row to form a sentence about a healthy habit. Write the healthy habit on the line below each row of word blocks.

1. | eight | sleep every night | Get | hours of |

___

2. | you sneeze | your mouth when | or cough | Cover |

___

3. | eyes, nose, | Keep your | hands | and mouth | away from your |

___

4. | with a | Eat a | variety of foods | well-balanced diet |

___

5. | regular | exercise | Get |

___

6. | after using | hands | and before eating | Wash your | the bathroom |

___

# The Chain Gang

You may have seen a cartoon in which a small fish gets eaten by a big fish. Then the big fish gets eaten by a bigger fish. On and on it goes. This feeding pattern is called a *food chain*. Energy is transferred from one organism to the other along the food chain.

The drawings below show the parts of a food chain, but they are not in order. Put the parts in order by drawing arrows from one object to another. Start with the source of energy. It's okay if the arrows cross one another.

## Web Master

Most organisms eat more than one kind of food. Therefore, an organism is usually part of many different food chains. Many different food chains interconnect to form a food web.

Use the organisms in the box to make a food web in the space below. You can either use the words or draw pictures. Use arrows to show how energy moves through the food web.

| coyote | deer | frog | grass | grasshopper | owl | rabbit | snake |
|---|---|---|---|---|---|---|---|

# Hidden Ecology

*Ecology* is the study of how organisms interact with the living and nonliving things in their environment. Two words about ecology are hidden in each puzzle below. The two words use all the letters in each puzzle. All the letters are in the correct order from left to right. Circle the letters that form the answer to clue A. The answer to clue B will be the word formed by the remaining letters. Write the words on the lines provided.

1. H A B I N I C T A T H E
   Clue A: the place where an organism lives

   ____________________

   Clue B: the role of an organism in its community

   ____________________

2. P R C O D O N U S U M C E E R R
   Clue A: includes organisms that make their own food

   ____________________

   Clue B: includes organisms that feed on other organisms

   ____________________

## Biome, My Home

A biome is a geographical region that has a particular climate and communities. There are several major biomes. Write the name of the biome next to its clue in the spaces below. The circled letters put in order will complete a mystery word that means all the places on Earth where life exists.

**Clues**

1. also called taiga; includes pine trees in the northern latitudes: (_) _ _ _ _ _ forest
2. biomes in water: _ _ _ _ _ (_) _
3. includes plains and prairies: _ _ _ _ (_) _ _ _ _

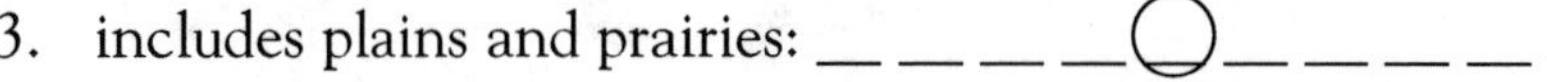

4. located near the equator; has a lot of rainfall: _ _ _ (_) _ _ _ _ rain forest
5. very dry; plants include cactus and sagebrush: _ _ _ (_) _ _
6. cold, dry area where few plants grow: _ _ _ _ (_) _

**Mystery Word:** _ _ O _ _ H _ _ E

# Different Kinds of Consumers

Most plants produce their own food by converting sunlight into the energy they need to live. Therefore, plants are producers. Most other organisms cannot use sunlight this way. They have to get their energy by eating, or consuming, other organisms. They are consumers.

Look at the three sets of words. The arrows show what the organism consumes. Decide which type of consumer is being shown, and fill in the lines to spell out the answer. The last part of each consumer name is filled in for you.

1. deer ⟶ grass, twigs, seeds

   __ __ __ __ __ vore

2. raccoon ⟶ crayfish, fruits, insects

   __ __ __ __ vore

3. lion ⟶ zebra, wildebeest, gazelle

   __ __ __ __ __ vore

# Up, Up, and Away

For thousands of years, the human population of the world was small. Recently, however, the population has increased dramatically, mostly in undeveloped or developing countries.

Use the Internet or other reference materials to answer the questions below. (A helpful source to consult is www.census.gov.)

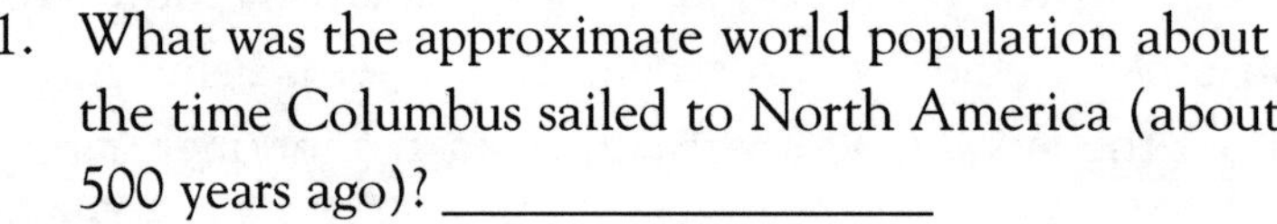

1. What was the approximate world population about the time Columbus sailed to North America (about 500 years ago)? ________________

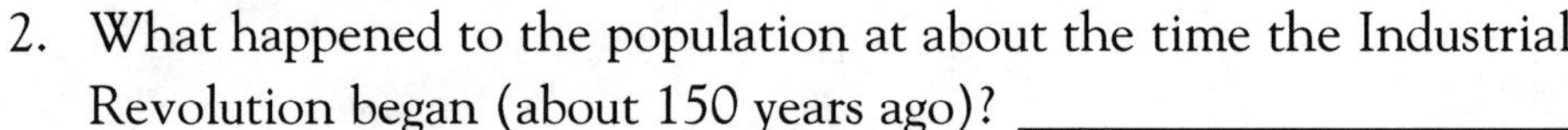

2. What happened to the population at about the time the Industrial Revolution began (about 150 years ago)? ____________________________

   ____________________________________________________________

3. What reasons can you give for this change in population growth?

   ____________________________________________________________

   ____________________________________________________________

4. Approximately what is the current world population? ________________

## What's the Matter?

You, this book, and any substance you can see and touch is matter. In fact, some things you cannot see are matter, too, such as the air. Matter is anything that has mass and takes up space.

Several words that have to do with matter are hidden in the "M" below. They may be forward or backward. Find and circle the words. Use the clues to help you.

**Clues**

1. a way to describe matter; a characteristic
2. solid, liquid, or gas
3. mass in a certain volume
4. tiny particle made of atoms

MOLECULEYTR
EPORPHL
OETATS
DENSITYAHLERT

# Adding Mass

One property of matter is mass. *Mass* is the amount of matter in something. You measure an object's mass in grams by using a balance. You can place standard masses, usually brass cylinders, on one pan until the other pan holding the object is balanced.

1. What is the mass of the apple below? Add up the grams shown on the standard masses to find out. ______

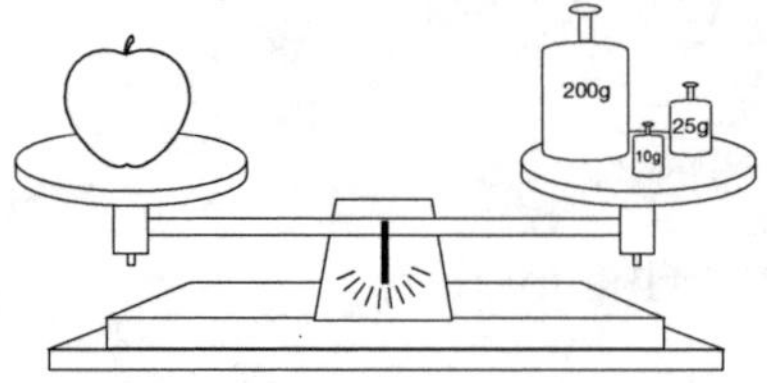


2. The rock on the balance has a mass of 145 g. Draw and label the correct standard masses on the balance to show this.

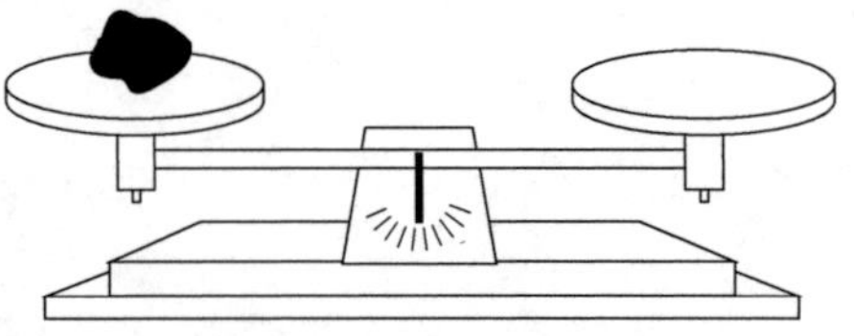

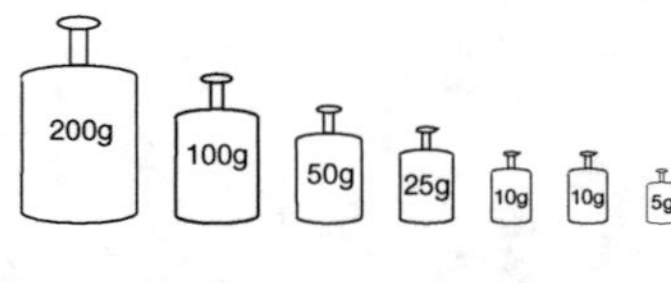

## Measuring Volume

You can use a graduated cylinder to measure the volume of a liquid in milliliters (mL). Usually the top of the liquid curves a little in the cylinder. Read the volume from the bottom of the curve.

Write the volumes shown in A and B. Then shade C and D to show the volumes 9.5 mL and 77 mL.

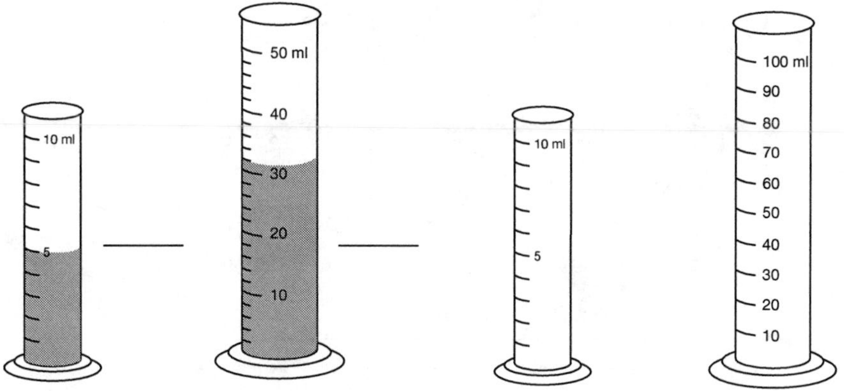

A ______  B ______  C  D

## Don't Be Dense

One of the most important properties of matter is density. *Density* is the amount of matter in a certain volume. It's a measure of how tightly a substance is packed into a volume. You can figure out a substance's density by dividing its mass by its volume (density = mass/volume). Density is measured in g/cm$^3$.

1. Suppose you have a piece of metal. Its mass is 49.72 g and its volume is 4.4 cm$^3$. What is its density?

   ______________

2. Another piece of metal has a mass of 33.9 g and its volume is 3 cm$^3$. What is its density? ______________

3. What can you say about the two pieces of metal? ______________

   ______________

## State of the Art

Matter can exist as a solid, a liquid, or a gas. Water, for example, can be ice (solid), liquid water (liquid), or water vapor (gas). The water particles behave differently in each state.

The squares below represent the states of matter. Use circles to represent the particles that make up the matter. Draw how the particles would look at each state. You might use arrows or other symbols in addition to circles.

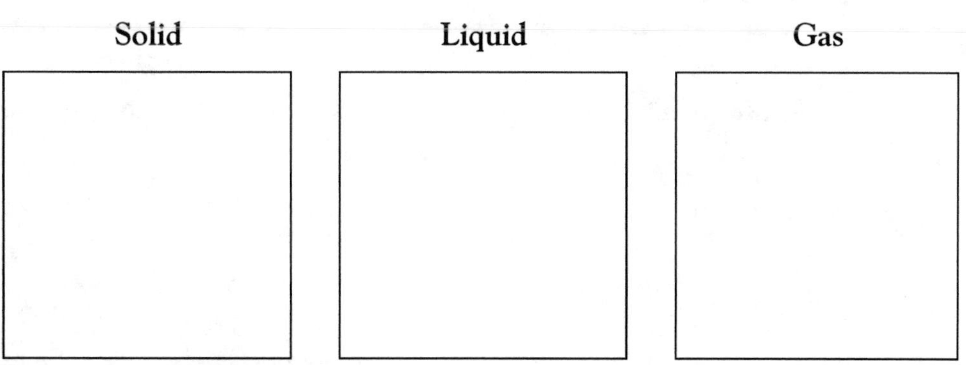

## Atomic Parts

Matter is made of tiny particles called *atoms*. Most matter is a combination of different kinds of atoms. Matter that contains only one kind of atom is an *element*.

Nearly every kind of atom has four parts. Find and trace the parts hidden in the puzzles. You can trace up, down, across, forward, or backward. Do not cross a line. A clue for the word is given below each puzzle.

1\.
n v u n
b s c w
u e l r
s p t g

clue: center of an atom

2\.
e l h e
g e x c
y c t r
s t n o

clue: negatively charged particle

3\.
a c w u
e t r o
c u a n
n e f h

clue: has no electrical charge

4\.
r n o n
e p t i
g r o s
i e j l

clue: positively charged particle

# Describe an Atom

Different atoms have different numbers of protons, neutrons, or electrons. Here are some important atomic relationships:

| | | |
|---|---|---|
| number of protons | = | number of electrons |
| atomic number | = | number of protons |
| mass number | = | number of protons + number of neutrons |

Answer the questions about the simplified model of a carbon atom.

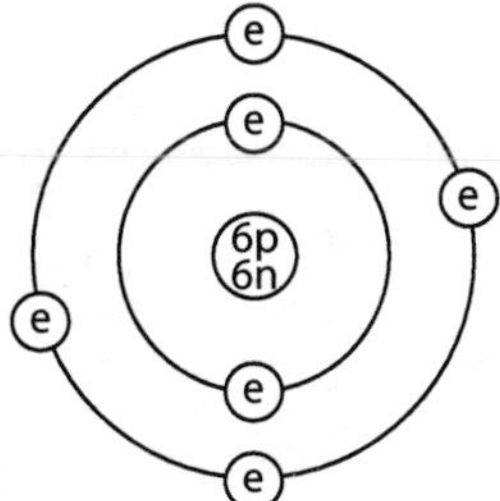


1. How many protons does it have? ________
2. How many electrons does it have? ________
3. What is its atomic number? ________
4. What is its mass number? ________

## The Same, But Different

All atoms of an element have the same number of protons, which is the atomic number of the element. But different atoms of the same element can have different numbers of neutrons. These different forms of the same element are called *isotopes* of the element.

Hydrogen has three isotopes. The most common isotope has one proton and no neutrons, as shown. Draw the other two isotopes.

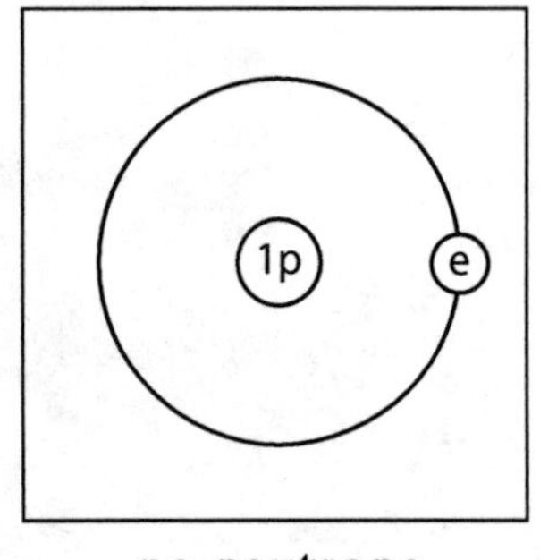


no neutrons

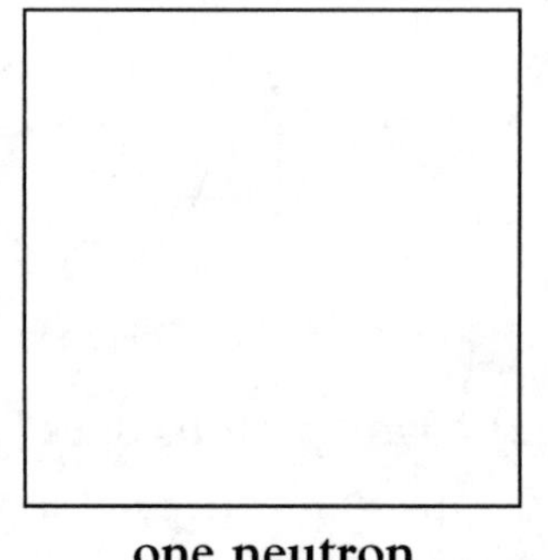

one neutron

two neutrons

## Strange Symbols?

There are more than 100 known elements. Each element has a symbol. Most of the symbols include the first letter of the element's name and sometimes a second letter from the name. But some symbols don't seem to make sense. Some of them are from Latin words.

The origins of some element names are given below. Write the name and symbol of each. Use a periodic table to help you.

1. from the Latin word *plumbum* because this element was used in ancient water pipes: ______________
2. from the Latin word *aurum*: ______________
3. from *natron*, which is a saltlike substance left behind when shallow lakes evaporated in Egypt: ______________
4. from the Latin word *argentum*, which means "white and shining": ______________
5. also known as *wolfram*: ______________
6. from *ferrum*, which is the Latin name for this element: ______________

# Element Code

The elements are arranged by atomic number in the periodic table. Read each statement below about the periodic table. In the puzzle, shade in the number of each true statement to reveal the symbol of the lightest element.

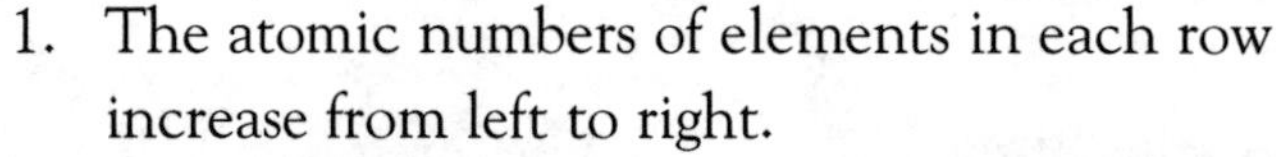

1. The atomic numbers of elements in each row increase from left to right.

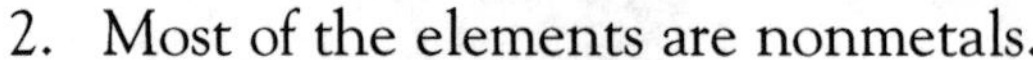

2. Most of the elements are nonmetals.
3. The number of electrons in an element increases in each row from left to right.
4. Elements in the same column are said to be in the same family.
5. Elements in the same column are very different from one another.

| 2 | 1 | 5 | 2 | 4 | 5 |
|---|---|---|---|---|---|
| 2 | 4 | 5 | 2 | 1 | 2 |
| 5 | 3 | 1 | 4 | 3 | 2 |
| 2 | 3 | 2 | 5 | 3 | 5 |
| 5 | 1 | 5 | 2 | 4 | 2 |

## Formula Fun

Most substances are made of more than one kind of element. Two or more elements combine to form a *compound*. For example, two atoms of hydrogen gas combine with one atom of oxygen gas to form water. So the chemical formula for water is $H_2O$.

Complete the table below by naming the number of elements in each compound.

| Compound | Formula | Number of Atoms |
|---|---|---|
| water | $H_2O$ | 2 hydrogen, 1 oxygen |
| salt | | |
| chalk | | |
| sugar | | |
| baking soda | | |

## Mixing It Up

Have you eaten a salad recently? How about spaghetti and meatballs? Maybe some iced tea? These are all *mixtures*.

The following pairs of terms are all about mixtures. Tell how the terms in each pair are alike and how they are different.

1. mixture/compound

   alike: ______________________________

   different: ______________________________

2. solution/suspension

   alike: ______________________________

   different: ______________________________

3. solvent/solute

   alike: ______________________________

   different: ______________________________

# Keys to Good Reactions

When two or more elements combine or rearrange, they undergo a *chemical reaction*. Scientists use a *chemical equation* to show a chemical reaction.

Use the chemical equation below to answer the questions that follow.

$$Cu + Cl_2 \longrightarrow CuCl_2$$

1. What are the reactants? ______________
2. What is the product? ______________
3. What does the arrow mean? ______________
4. How many copper atoms are on each side of the equation? ______________
5. How many chlorine atoms are on each side of the equation? ______________

## A Balancing Act

In a chemical reaction, mass is neither created nor destroyed. Therefore, a chemical equation must be balanced. A balanced equation shows that the number of atoms before the reaction is the same as the number after the reaction.

Look at each equation below. If it is balanced, write *balanced* in the space. If it is not balanced, add coefficients to balance it.

1. $S + O_2 \longrightarrow SO_2$ ____________

2. $H_2 + O_2 \longrightarrow H_2O$ ____________

3. $C_6H_{12}O_6 \longrightarrow C + H_2O$ ____________

# All Kinds of Reactions

Chemical reactions produce new substances with properties that are different from those of the original substances. There are different kinds of chemical reactions. Use the clues below to name some.

| | Clue | Reaction |
|---|---|---|
| 1. | chemical reaction that gives off energy | __ __ __ thermic |
| 2. | chemical reaction that absorbs energy | __ __ __ __ thermic |
| 3. | one element replaces another element | __ __ __ __ __ __- replacement |
| 4. | elements in two compounds switch | __ __ __ __ __ __- replacement |

## pH Sudoku

In a sudoku puzzle, each three-by-three box and each row and column in a nine-by-nine grid has to have the numbers 1 through 9. You can play sudoku with words, too. Solve the puzzle for the words *neutral pH*. Do not leave a space between the two words. The middle box is done for you.

| | | | | | | | | |
|---|---|---|---|---|---|---|---|---|
| | R | L | U | P | | | | A |
| | | N | | | L | | R | |
| E | | | | H | | P | | T |
| | N | E | P | T | U | | H | |
| A | | | R | N | H | | | L |
| | U | | L | E | A | T | P | |
| R | | H | | U | | | | P |
| | E | | A | | | R | | |
| N | | | | R | P | L | U | |

## Acid or Base?

Acids and bases are two important groups of compounds. You use them every day, even though you may not realize it.

For each item below, write A if it is a property or an example of an acid. Write B if it is a property or an example of a base. *Hint:* There are five of each.

1. soap ____
2. baking soda solution ____
3. tastes sour ____
4. includes lemon juice ____
5. tastes bitter ____
6. produced in stomach to digest food ____
7. has a pH of 4 ____
8. feels slippery ____
9. turns blue litmus paper red ____
10. has a pH of 10 ____

## How Fast?

When you tell how fast an object is moving, you give its speed. The formula for speed is: speed = distance/time.

If a car traveled 30 miles in one hour, its speed was 30 miles/hour, or 30 miles per hour. This is actually its average speed. It traveled slower or faster at different times during the trip. Calculate the average speed in each problem below.

1. Corrine swims 100 meters in 50 seconds. What is her average speed? ________________

2. What is the average speed of a car that travels 300 miles in 6 hours?

   ________________

3. Joe rides his bike to a friend's house 3.5 miles away. It takes him 15 minutes (0.25 hour) to get there. What was his average speed in miles/minute? in miles/hour? ________________

## Graphing Motion

You can describe an object's motion with a graph. The table shows how far three cars traveled in 4 hours. Plot the distance traveled by each car. Then answer the question.

| Elapsed Time in Hours | Distance Traveled in Miles | | |
|---|---|---|---|
| | Car A | Car B | Car C |
| 1 | 50 | 45 | 30 |
| 2 | 100 | 90 | 60 |
| 3 | 150 | 135 | 90 |
| 4 | 200 | 180 | 120 |

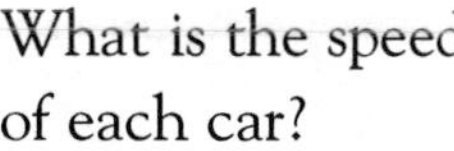

What is the speed of each car?

A. ______

B. ______

C. ______

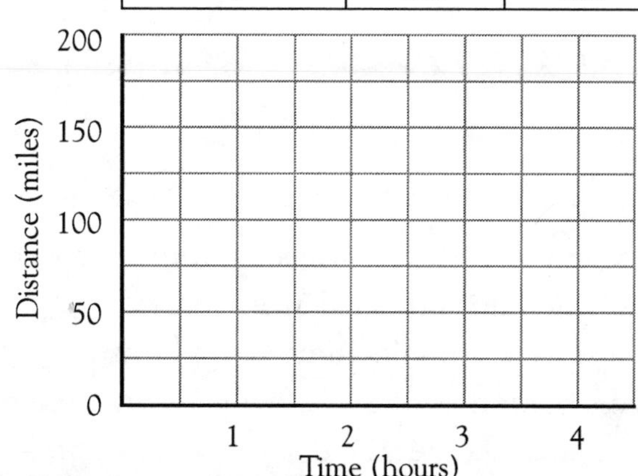

Daily Warm-Ups: General Science

# Combining Forces

A force is a push or a pull. Arrows show the strength and direction of forces.

The sets of arrows show three situations involving forces. Answer the questions about the situations.

A 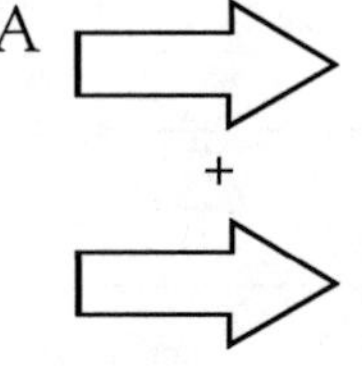

B 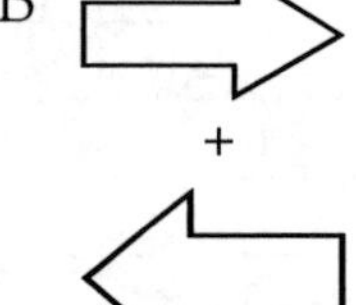 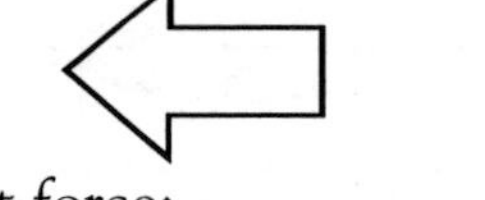

C 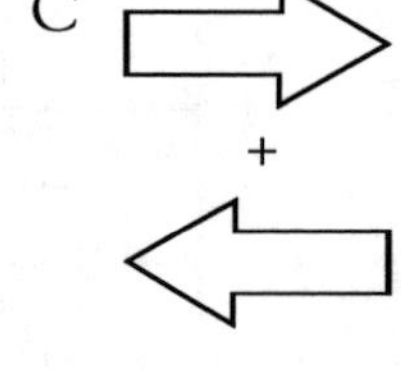

net force:  net force: net force: 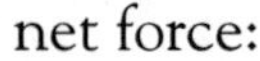

1. Which situation shows two equal forces exerted in the same direction? ____
2. Which situation shows balanced forces? ____
3. Which situation results in no net motion? ____
4. Which situation results in the greatest net force? ____
5. Draw or state the net force of each situation.

# Obey the Law (of Motion)

In the 1600s, Sir Isaac Newton developed the *laws of motion*. These three laws explain all motion in the universe, from the beating of the heart to the movements of planets.

The list below includes examples and descriptions of the three laws of motion. Classify each item by writing 1, 2, or 3 on the line to indicate the first, second, or third law of motion.

____ 1. Every action has an equal and opposite reaction.

____ 2. An object at rest remains at rest unless a force acts on it.

____ 3. A frog jumping off a lily pad pushes the pad backward.

____ 4. A full cart of groceries is harder to push than an empty cart.

____ 5. force = mass × acceleration

____ 6. When you walk, your feet push the ground and the ground pushes your feet.

____ 7. also called the "law of inertia"

____ 8. A spacecraft has to fire rockets to change its direction when traveling in space.

# Gravity Sudoku

A ______ circles Earth in space because gravity constantly pulls it down while its inertia keeps it moving forward.

To discover the missing word in the sentence above, unscramble the nine letters in the center of the sudoku puzzle below. Then fill in the puzzle so that each nine-by-nine square, each row, and each column use all the letters only once.

| | | | | | | | | |
|---|---|---|---|---|---|---|---|---|
| I | | | S | | | | L | T |
| | | T | A | L | | S | | E |
| S | L | | | L | | | | I |
| | | | E | A | L | I | | |
| E | E | I | L | S | T | L | A | T |
| | | S | I | E | T | | | |
| T | | | | E | | | I | |
| A | | L | | T | S | E | | |
| L | E | | T | | | | S | L |

# Gravity Quiz

When you throw a ball in the air, you know it will come back down. The force of gravity pulls it down. Gravity acts between any two objects, such as the ball and Earth. Show what you know about gravity by answering the following questions.

1. A bowling ball, a volleyball, and a basketball are dropped at the same time. In what order will the balls hit the ground?

   ______________________________________________

2. A baseball, a tennis ball, and a sheet of paper are dropped at the same time. In what order will the objects hit the ground?

   ______________________________________________

3. Fill in the vowels to name the kind of friction that makes one of the objects in number 2 fall more slowly than the others.

   __ __ r r __ s __ s t __ n c __

4. The strength of gravity between two objects depends on the ____________ of the objects and the ____________ between the objects.

5. What famous scientist discovered the law of gravity? ____________

## A Hidden Formula

By now, you probably know that some words have different meanings in science than they do in everyday use. Take the word *work*, for example. Work can mean many things. It might mean "thinking a lot." It could mean "anything that makes you sweat." Maybe it is anything you don't like to do. In science, though, work is done only when a force causes an object to move a distance.

This idea of work is expressed in a formula. That formula is spelled out in the puzzle below. Darken the letters b, g, h, p, and y to reveal the formula. Then write it on the line provided.

94

| b | w | y | o | h | r | k | p | p | e | g | b | q | h | u | a | h | l |
|---|---|---|---|---|---|---|---|---|---|---|---|---|---|---|---|---|---|
| s | p | b | h | f | b | o | y | r | p | c | e | p | y | t | p | i | g |
| g | m | e | y | h | s | p | d | i | b | s | t | g | a | n | h | c | e |

Formula: ______________________________

## How Does It Work?

A machine is a tool or device that makes work easier to do. A machine also lets us do tasks that would be extremely difficult or impossible to do without the machine. A machine could be as simple as a stick or more complex than an airplane.

Find out how machines work by unscrambling the words in the boxes to complete the paragraph. Write the words on the spaces provided.

When you use a machine, you apply a force to the machine, called the [UTNIP] ________ force, or effort force. The force exerted by the machine is the [PTUUTO] ________ force, or resistance force. A machine makes work easier in three ways. It either changes the [NTUOAM] ________ of force you apply, the [COIDIRNTE] ________ of the force, or the [ECIDNAST] ________ over which you apply the force.

## Machine Magic

There are six basic kinds of machines that have no more than a few parts. They are called *simple machines*. Match each simple machine or related term in column A with its description in column B. Write the number in the box of the matching letter. To discover the magic number, add the numbers in a row, column, or diagonal. If all your answers are correct, the number will be the same.

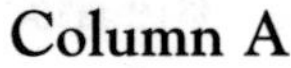

**Column A**

A. gear
B. pulley
C. wheel and axle
D. lever
E. inclined plane
F. fulcrum
G. wedge
H. load
I. force

**Column B**

1. an object to be moved
2. wheel with teeth
3. fixed point around which a lever pivots
4. example: doorknob
5. ramp
6. example: ax head
7. bar that pivots around a fixed point
8. push or pull
9. grooved wheel with rope set inside it

| A | B | C |
|---|---|---|
| D | E | F |
| G | H | I |

## Seesaw Science

There are three kinds, or classes, of *levers*. The table describes them. Look at the list of levers below. Write each under the correct class in the table. *Hint:* Each class has three examples.

| baseball bat | crowbar | scissors |
|---|---|---|
| bottle opener | paper cutter | seesaw |
| broom | rake | wheelbarrow |

| Class 1 Lever | Class 2 Lever | Class 3 Lever |
|---|---|---|
| Fulcrum is between effort force and resistance force. | Resistance force is between effort force and fulcrum. | Effort force is between resistance force and fulcrum. |
| 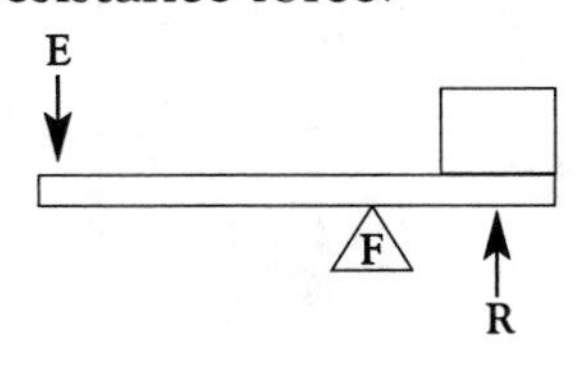  | 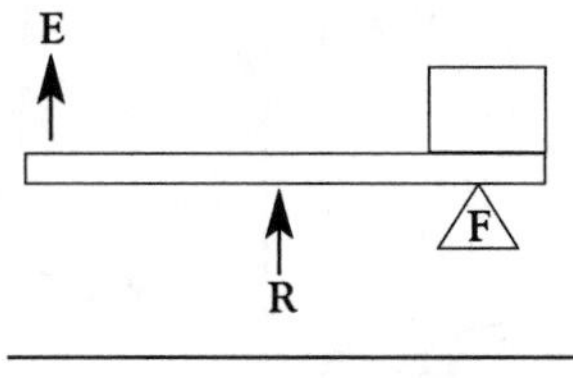  | 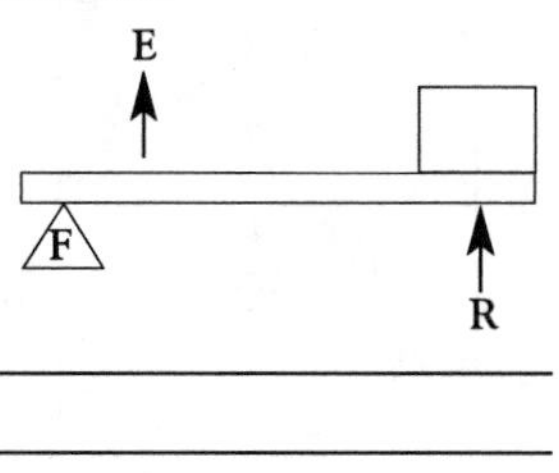  |
| ______ | ______ | ______ |
| ______ | ______ | ______ |
| ______ | ______ | ______ |

## One, Two, Three, Pull!

A *pulley* is a grooved wheel with a rope set inside it. By pulling on the rope, you change the amount or direction of the input force. The number of times a machine, such as a pulley, multiplies the input force is the machine's *mechanical advantage* (MA). The higher the MA, the less force you need to use. In a pulley system, the MA equals the number of rope sections that support, or pull up on, the pulley. Sections of rope that pull down on the pulley system do not add to the MA. Name the MA for each pulley system below. Arrows show the input force. The MA will either be 1, 2, or 3.

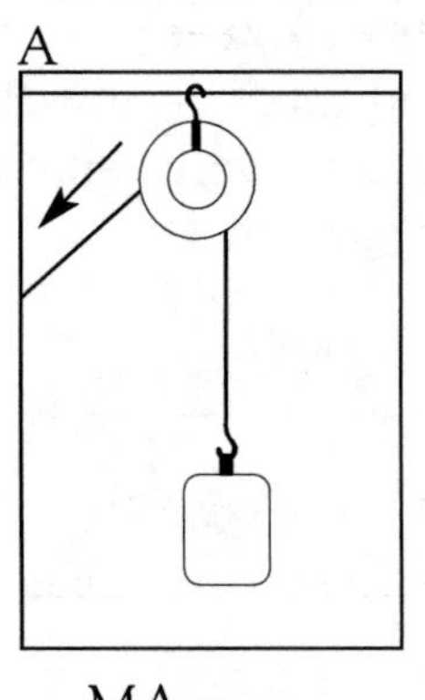


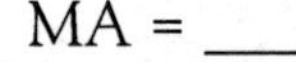


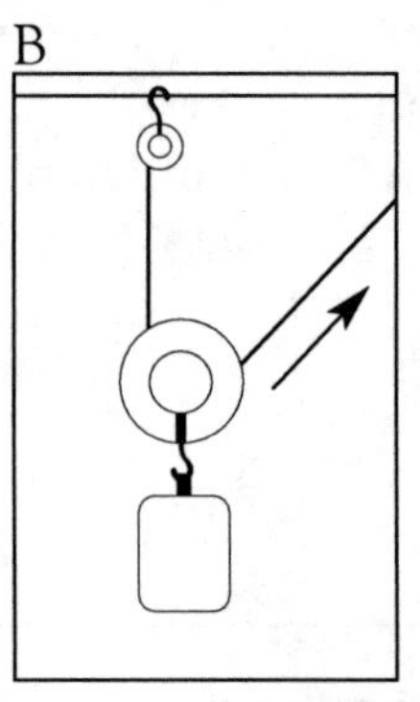


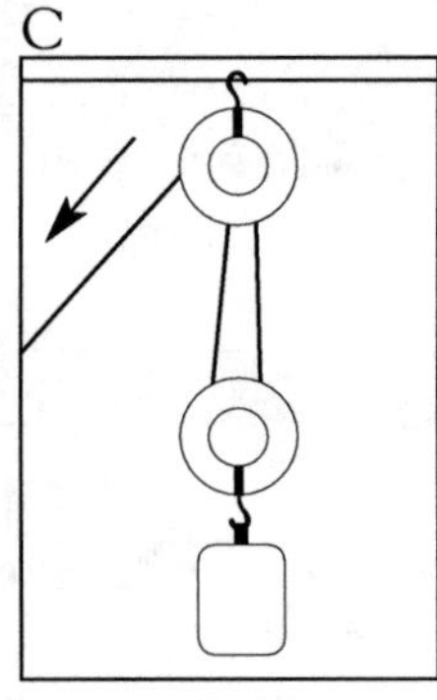


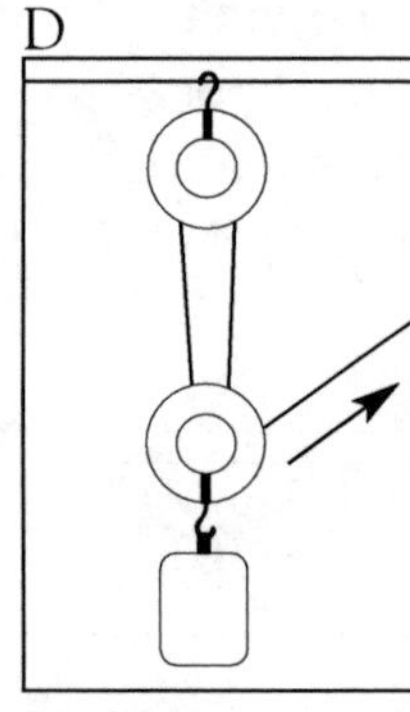

## Choose Your Machine

You know about the six kinds of simple machines. Do you know when to use them? For each situation below, decide which simple machine would be most helpful. Write the machine on the line provided.

1. Two wooden boards are stuck together. I need to separate them with a(n) ________________.
2. We have to put a heavy box into the back of a pickup truck. We'll put it on a dolly and roll the dolly up a(n) ________________.
3. To keep our food away from bears while camping, we have to hang our sack of food over a tree branch. The branch acts like a(n) ________________.
4. The leg is coming loose from the chair. I'd better use a(n) ________________ to hold the pieces together.
5. I have to use a screwdriver to get the lid off a paint can. I'll use the screwdriver as a ________________ lever. When I mix the paint, I'll use a wooden ruler as a ________________ lever.

## Compound Machines

A *compound machine* is a machine that uses two or more simple machines. Most machines are compound machines. For example, the handle of a crowbar is a lever. But the narrowed, tapered end of a crowbar is a wedge.

See if you can name the simple machines that make up each compound machine in the chart.

| Compound Machine | Simple Machines |
|---|---|
| scissors | |
| fork | |
| hand-held can opener | |
| bicycle | |

# Mapping Energy

*Energy* is the ability to do work. Without energy, there would be no light and no heat, and nothing would move. Fill in the concept map below to get a better understanding of energy and some of the many forms it can take.

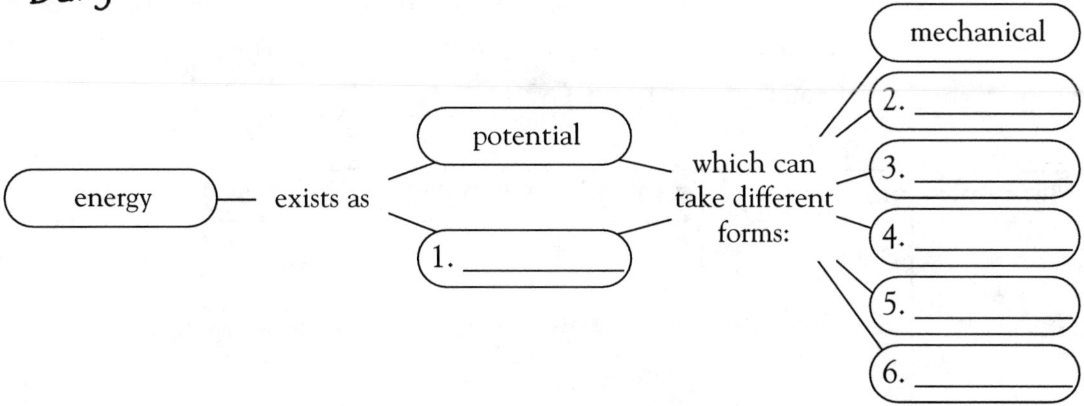

energy — exists as
- potential
- 1. ________

which can take different forms:
- mechanical
- 2. ________
- 3. ________
- 4. ________
- 5. ________
- 6. ________

Daily Warm-Ups: General Science

# A World of Energy

Energy is all around you. It's inside you, too. Your body uses the chemical energy in food to perform all the functions that keep you alive.

The objects and actions below are a few examples of energy in use. Write each example in the correct category in the chart. Some may go in more than one category.

| | | | |
|---|---|---|---|
| wind | kicking a ball | moving electrons | fire |
| lightning | boiling water | gasoline | battery |
| the sun | head of a match | | |

| Mechanical | Heat | Light | Chemical | Electrical | Nuclear |
|---|---|---|---|---|---|
| | | | | | |

# Changing Energy

Energy cannot be created or destroyed, but it can change from one form to another. For example, the electrical energy traveling through wires changes to light energy in a light bulb.

Below are several examples of using energy. Decide what energy changes are taking place, and fill in the lines.

1. A toaster heats bread to make toast. _______________ energy changes to _______________ energy to heat up coils in the toaster.
2. You get energy from food. Your body changes _______________ energy in food into _______________ energy to move your muscles.
3. A battery is used to operate the hands of a clock. Chemical energy in the battery changes to _______________ energy in a motor. That energy changes to _______________ energy as the hands move.

## Drop-Letter Energy Puzzle

You know that energy cannot be created or destroyed. So when one form of energy changes to another form, no energy is destroyed in the process.

This rule about energy has a name. Unscramble the letters below to find it. Place one letter from each column of letters in each space above that column. Use the letters only once in each column of spaces. The filled-in spaces separate words. This puzzle has six words.

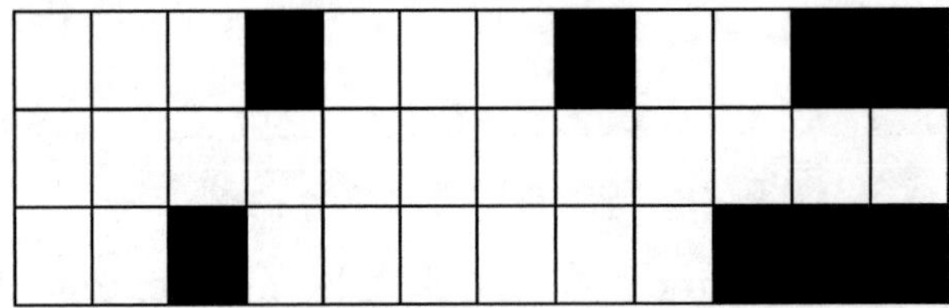

| O | F | E | S | N | E | V | G | T | F | O | N |
|---|---|---|---|---|---|---|---|---|---|---|---|
| T | O | N | E | E | R | W | A | Y | I | | |
| C | H | | | L | A | R | | O | | | |

## More Power

Suppose you and a friend rake leaves in your backyard. You each rake exactly half the yard. The leaves are spread evenly over the whole yard. Assume you both moved the rake the same number of strokes. But you finish your half in 30 minutes while your friend takes 40 minutes. Answer the questions below. To help you, analyze this formula: power = work/time.

1. Who did more work? Explain your answer.

_______________________________________________

_______________________________________________

_______________________________________________

2. Who used more power? Explain your answer.

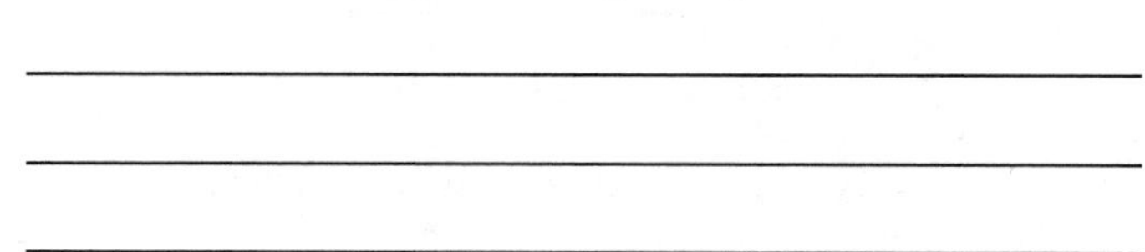

# Things Are Heating Up

All of the words in the scrambles below have to do with heat energy. Use the clues to unscramble the letters and form the words. Then unscramble the circled letters to find the mystery word.

1. the transfer of energy by waves through space

   T D A N R O I A I

2. the transfer of energy through direct contact between particles

   C D T O U I O C N N

3. the transfer of energy by the movement of currents in a gas or liquid

   V N T E O C N O C I

4. temperature scale on which water freezes at 0°

   L S S U E I C

Mystery Word: a unit for measuring heat

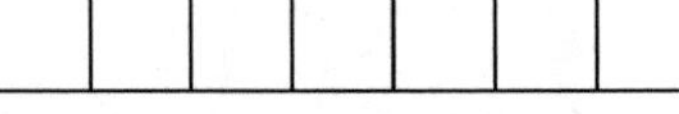

## Heat Sudoku

A(n) ______ does not conduct heat well. It slows the movement of heat.

To discover the missing word above, unscramble the nine letters in the center of the sudoku puzzle below. Then fill in the puzzle so that each nine-by-nine square, each row, and each column use all the letters only once.

|   | R |   | S |   | I |   | L |   |
|---|---|---|---|---|---|---|---|---|
|   |   | A | R |   | N | I |   |   |
| L | I |   |   |   |   |   | R | A |
|   |   | T | U | I | S | R |   |   |
| N | U |   | A | O | L |   | T | I |
|   |   | O | N | T | R | U |   | L |
| T | O |   |   | R |   |   | S | N |
|   |   | L | O |   | U | A |   |   |
|   | A |   | L |   | T |   | U |   |

## Design an Experiment

Suppose you want to build a container that will keep hot liquids hot. You decide to try an experiment to find the best materials to use for insulation. Use the materials below to write an experimental procedure that will help you find the best insulation.

**Materials**
plastic cups
coffee cans
hot water
thermometers
clock
insulating materials such as plastic foam, newspaper, cork, rubber, and so forth

______________________________________________

______________________________________________

______________________________________________

______________________________________________

# Catch a Wave

Waves carry energy from place to place. For example, you can see and hear because light and sound energy travel in waves from a source to your eyes and ears.

Not all waves are alike. They have different heights and lengths.

Label the wave diagram using the words below. Add any lines necessary to the diagram.

| amplitude<br>crest<br>trough<br>wavelength |
|---|

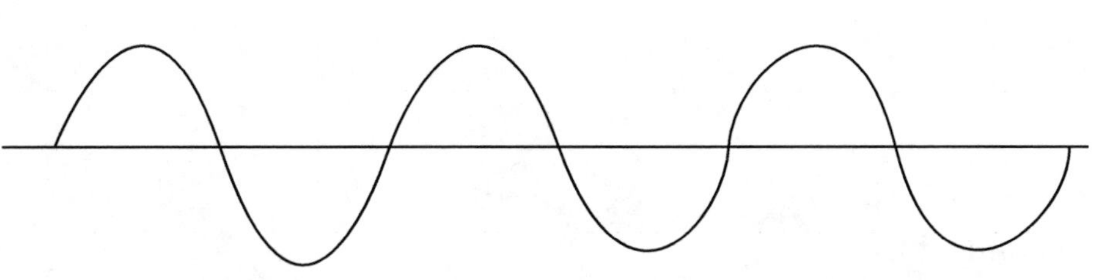

## Light It Up!

When light hits an object, it can do one of four things. It can bend, bounce off the object, pass through the object, or get absorbed by the object. The theme of the following crossword puzzle is about how light acts. Use the clues to solve the puzzle.

**Across**

3. light passes through
5. bouncing of light off an object

**Down**

1. breaks up light into its colors
2. blocks light completely
4. only some light passes through
6. bending of light

## Color My World

A prism separates white light, such as sunlight, into its many colors. This separation of colors is called the *visible spectrum*. You see the visible spectrum when you look at a rainbow. Water drops in the air act like tiny prisms and separate sunlight into the colors of the rainbow.

Label or color the rainbow below to show the colors and order of the visible spectrum. *Hint:* The first letters of the colors make the name ROY G. BIV.

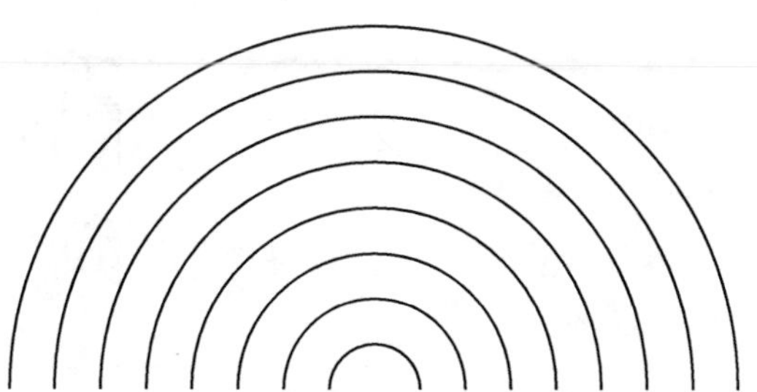

# Sounds of Silence

The intensity, or strength, of sound is measured in units called *decibels*. Every increase of 10 decibels means the sound's intensity is ten times greater. Sounds greater than 120 decibels are painful to most people.

Below is a decibel scale. Think about the sounds listed. Then try to place the sounds in the correct order on the scale.

**Sounds**

car horn
jet engine nearby
library
normal conversation
thunder overhead
whisper

**Decibels**

| Sound | Decibels |
|---|---|
| | 150 |
| ____________ | 140 |
| | 130 |
| | 120 |
| ____________ | 110 |
| | 100 |
| ____________ | 90 |
| | 80 |
| | 70 |
| ____________ | 60 |
| | 50 |
| ____________ | 40 |
| | 30 |
| | 20 |
| ____________ | 10 |

## Mystery Ship

Sound is made when matter vibrates. The vibrations move out in all directions as sound waves. Sound waves travel through solids, liquids, and gases.

Look at the drawing below. Explain what it shows. Use the terms *sound waves*, *reflect*, *send*, *receiver*, and *sonar*.

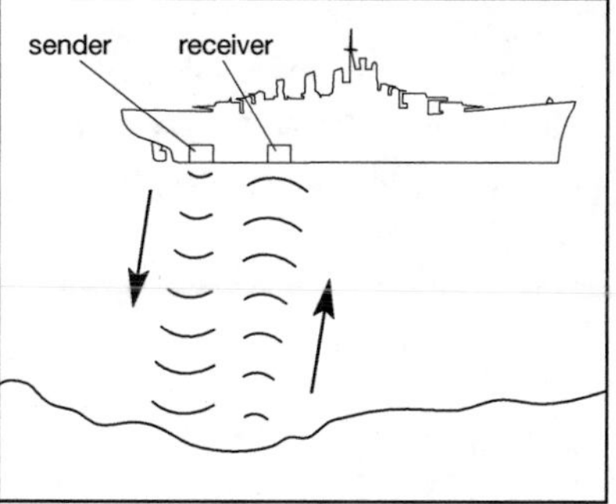


______________________________________________

______________________________________________

______________________________________________

______________________________________________

# What a Shock!

The protons and electrons in atoms have electric charge. Protons have a positive charge (+), and electrons have a negative charge (–). Particles with the same charge repel each other. Particles with different charges attract each other. Electrons move easily from one material to another. A buildup of electrons gives an object a negative charge.

You have probably gotten a shock at some time after walking across a carpet and then touching a metal doorknob. Show why this happens by completing the drawings below. The circles represent particles. Draw a minus sign or a plus sign in each circle to show what you think the charge should be.

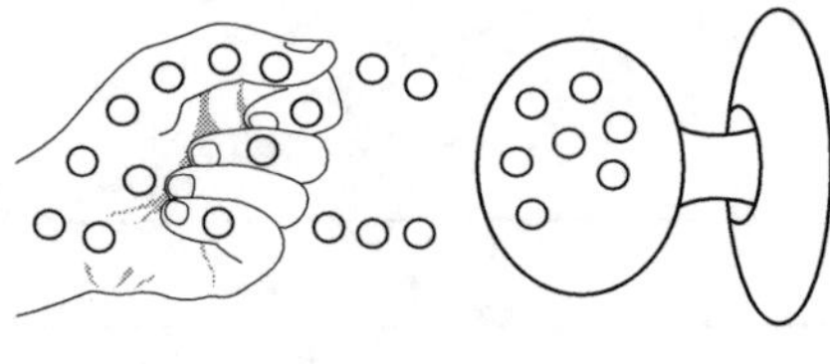

# Follow the Path

A continuous flow of electric charge is an *electric current*. Current flows only if the path is unbroken. This path is called a *closed circuit*. There are two main kinds of closed circuits. Show each kind by completing the drawings below. Add the wires that connect the lightbulbs to the battery. Then name each kind of circuit.

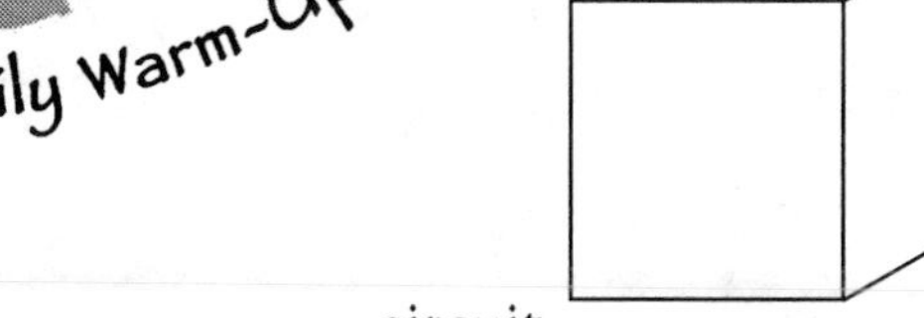

____________ circuit

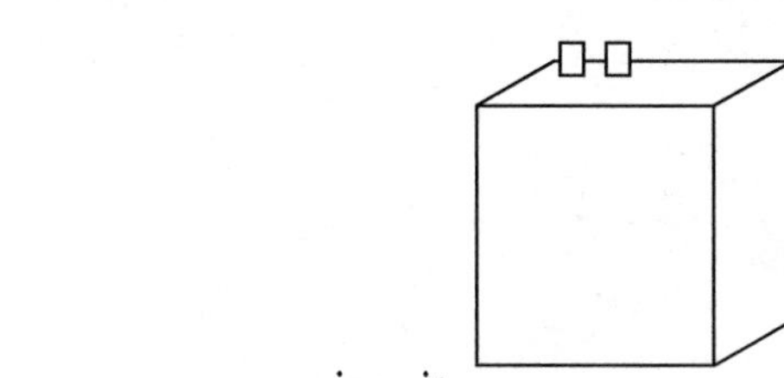

____________ circuit

# It's Electrifying!

It is difficult to imagine living without electricity. It powers many of the machines we use every day.

Each sentence below is about electricity. Complete each sentence by circling the letters that spell the missing word.

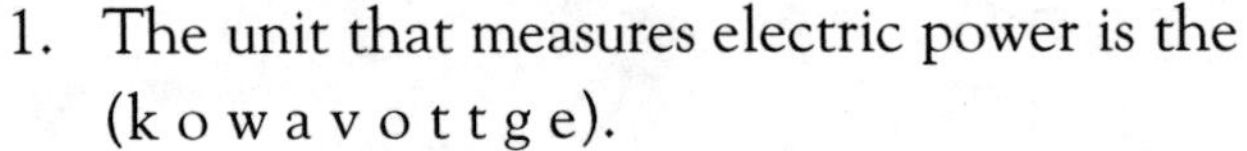

1. The unit that measures electric power is the (k o w a v o t t g e).
2. A conductor has low (o c r e s e l i s i n t a n o r c e), so electric current flows easily through it.
3. When a light switch is off, the circuit is broken, or (c s p o p w e h n).
4. A device that raises or lowers voltage is a (t h r o a l n s f w o r z a m e r).

# The Meter is Running

Buildings have electric meters that tell the electric company how much electricity was used for the month. The meter measures energy use in kilowatt-hours (kWh). Each dial stands for a place value. The dial farthest right has a place value in ones, the second in tens, the third in hundreds, and the fourth in thousands. To determine the electricity used in a month, you subtract last month's total from the current total.

Look at the meters below.

**July 15**

_____ _____ _____ _____

**August 15**

_____ _____ _____ _____

1. How much electricity was used at each reading? ____________
2. How much electricity was used during the month? ____________
3. If it costs $0.15 per kWh, what is this month's electric bill? ____________

## I'm Attracted to You

A magnet is an object that attracts certain materials. The force that attracts these materials is called *magnetism*.

In the word search below, find five materials written across that are attracted to a magnet. Find five materials written up and down that are not attracted to a magnet. The words can read forward or backward.

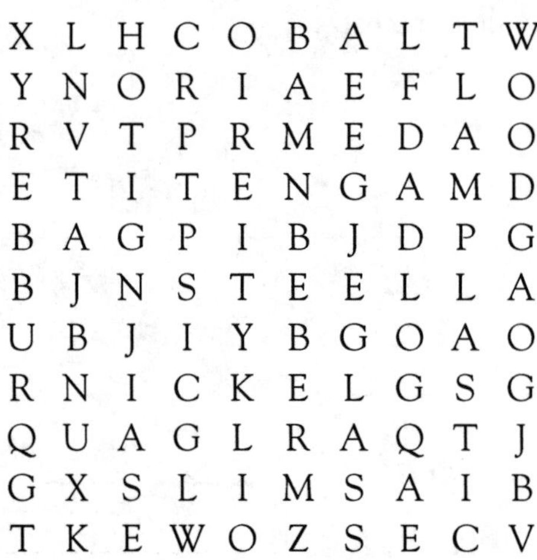

X L H C O B A L T W
Y N O R I A E F L O
R V T P R M E D A O
E T I T E N G A M D
B A G P I B J D P G
B J N S T E E L L A
U B J I Y B G O A O
R N I C K E L G S G
Q U A G L R A Q T J
G X S L I M S A I B
T K E W O Z S E C V

# Magnetic Password

Each column below includes a series of words. Read through each series, one word at a time, to figure out the mystery word. Each mystery word has something to do with magnetism.

1. magnets
   coil
   wire
   rotate
   electric current

**Mystery Word:**

__ __ __ __ __ __ __ __ __

2. Earth
   North Pole
   South Pole
   magnetic field
   instrument

**Mystery Word:**

__ __ __ __ __ __ __

3. electricity
   coil
   electromagnet
   reverse
   mechanical energy

**Mystery Word:**

__ __ __ __ __

# High Tech

An electronic device uses electric current to represent coded information. You probably use several electronic devices throughout the day—radio, television, CD player, game systems, cell phones, and many more.

Test your knowledge of electronics with this matching quiz. Write the letter of the word in column B that matches the definition in column A.

**Column A**

___ 1. temporary storage area for data in a computer

___ 2. chip that contains many electronic components

___ 3. system in which information is represented by two digits

___ 4. electronic machine that stores, processes, and retrieves information

___ 5. international computer network

**Column B**

a. binary code

b. computer

c. integrated circuit

d. Internet

e. RAM

# Map Puzzle

A map is a model of Earth's surface. You use maps to find places. You might also use maps to see borders, such as state borders, and to see how something, such as rainfall, is spread over an area.

Find the answers to the clues in the puzzle below. Words can read forward, backward, up, down, or diagonally. Then unscramble the circled letters to spell the name for a book of maps.

1. instrument used to tell direction
2. opposite of north
3. circles Earth halfway between North and South poles
4. opposite of west
5. distance north or south of equator

| | | | | | | | | | | | | |
|---|---|---|---|---|---|---|---|---|---|---|---|---|
| A | S | I | U | B | T | K | W | D | E | U | H | P |
| D | B | E | D | U | T | I | (T) | A | (L) | D | T | H |
| I | C | O | M | P | (A) | S | S | B | V | A | U | I |
| C | O | F | E | R | O | T | (A) | U | Q | E | O | A |
| J | T | A | C | X | M | O | E | S | M | R | (S) | T |
| Q | E | L | P | B | S | D | R | C | L | G | S | M |

A book of maps is a(n) __ __ __ __ __.

## Where Am I?

You can locate any place on Earth if you know its latitude and longitude. *Latitude* is the distance north or south of the equator. *Longitude* is the distance east or west of the prime meridian. Latitude and longitude are measured in degrees.

Use the locations to tell whether each statement is true or false. Write *true* or *false* on the line provided.

**Locations**

Chicago: 42°N, 88°W

Sydney: 34°S, 151°E

Denver: 40°N, 105°W

Paris: 49°N, 2°E

_______ 1. Denver is farther west than Chicago.

_______ 2. Paris is farther north than Denver.

_______ 3. Paris is in the Southern Hemisphere.

_______ 4. Chicago is closer to the equator than Denver is.

_______ 5. Paris is only two degrees west of the prime meridian.

## What Time Is It?

The map shows time zones for most of the United States. Each time zone is one hour earlier than the zone east of it. It is noon in New York City. Draw the clock hands to show the time in all four time zones. Then answer the following questions.

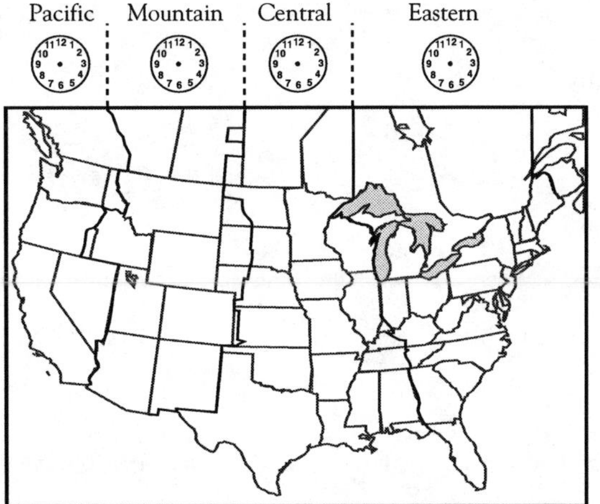


1. Why is each time zone to the east later than the time zone to the west?

_______________________________________________

2. Into how many time zones is Earth divided? ________

# Topographic Maps 1

A *topographic map* shows changes in elevation on Earth's surface. The key to understanding a topographic map is understanding contour lines. A *contour line* connects points of equal elevation. The vertical distance between two successive contour lines is the *contour interval*.

Use the simple topographic map below to answer the questions. The elevations are marked in feet.

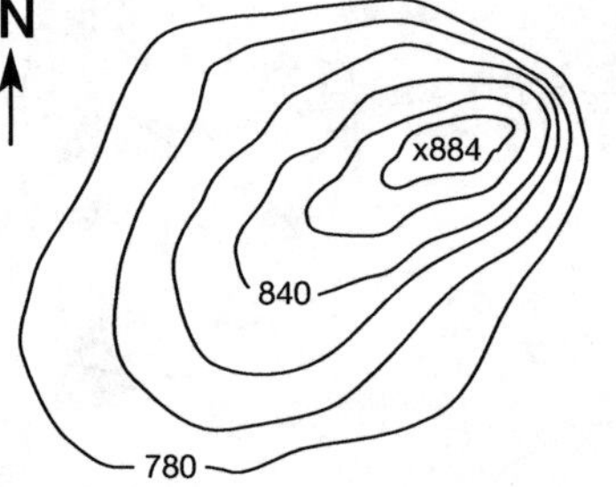


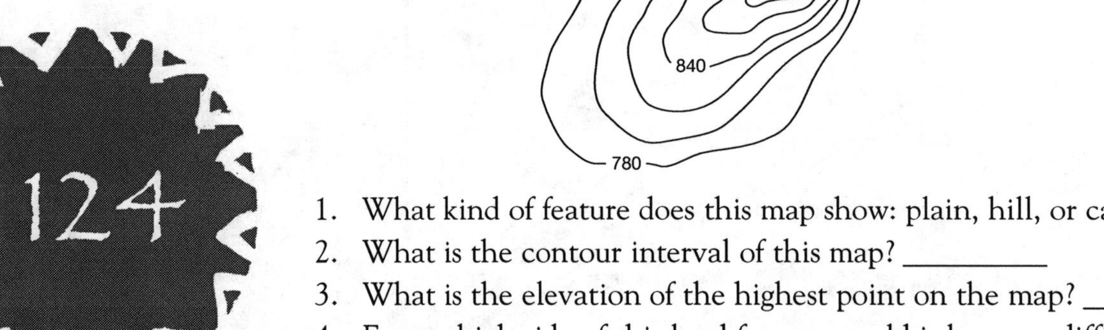


1. What kind of feature does this map show: plain, hill, or canyon? __________
2. What is the contour interval of this map? __________
3. What is the elevation of the highest point on the map? __________
4. From which side of this land feature would it be most difficult to reach the highest point? Why? __________

# Topographic Maps 2

Look at the topographic map below. The elevations are in feet.

Look at the line from A to B. Draw a profile, or side view, of how the landscape would look along this line. Label points A and B on your profile.

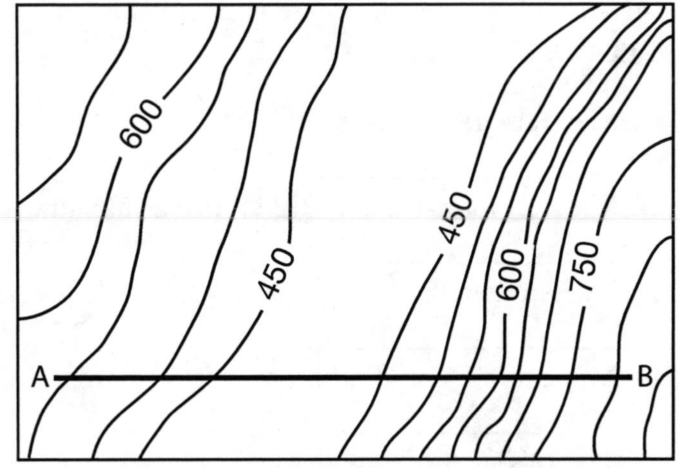

# What Makes a Mineral?

Many of the items around you contain one or more *minerals*. From the lead in your pencil to some of the ingredients in your food, minerals are an important part of your life.

In order to be a mineral, a material has to have five characteristics. Unscramble the words in capital letters in the sentences below to find out these characteristics.

1. Minerals are DOSSLI.

    ___________________________

2. Minerals form TAYNARLLU, not artificially.

    ___________________________

3. Each mineral has a consistent MCEHLACI composition.

    ___________________________

4. Minerals are not VILAE or made of living things.

    ___________________________

5. The MSTOA of minerals are arranged in definite, repeating patterns.

    ___________________________

## What's the Difference?

Sometimes it's difficult to tell what is a mineral and what is not. A good way to tell the difference between two things is to use a Venn diagram. Look at the Venn diagram below. It shows what is the same and what is different between a mineral, galena, and a non-mineral, concrete. The materials share two of the five characteristics of a mineral. But only the mineral has all five characteristics. Fill in the Venn diagram.

| Galena | Both | Concrete |
|---|---|---|
| forms naturally | 1. ________ | made by people |
| consistent chemical composition | 2. ________ | chemical composition varies |
| atoms in repeating patterns | | no repeating atomic pattern |

## Hidden Minerals

There are about 3000 known minerals. However, only about 20 of them make up 90 percent of the rocks in Earth's crust. Eight of these minerals are listed below. Write the missing letters in the spaces provided to complete the name for each mineral.

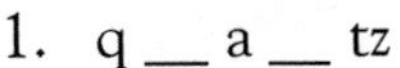

1. q __ a __ tz
2. __ el __ sp __ r
3. m __ c __
4. ca __ ci __ e
5. do __ o __ ite
6. __ al __ te
7. gy __ s __ m
8. ol __ vi __ e

## Mystery Mineral

One way to identify a mineral is by its hardness. A mineral that scratches an object is harder than the object. Also, if mineral X scratches mineral Y, then mineral X is harder than mineral Y.

Read the statements below. Then figure out the order of hardness for the four minerals.

Mineral A scratches mineral B.

Mineral C scratches mineral A.

Mineral D scratches mineral A but not mineral C.

**Hardest mineral** **Softest mineral**

____ ____ ____ ____

## You're a Gem!

Gems are minerals that are prized for their beauty and rarity. Each month is represented by a certain gem, called its birthstone.

Unscramble each group of letters below to discover each birthstone.

1. January: T E N R A G ____________________
2. February: Y S T A M H T E ____________________
3. March: A Q M A U A R I N E ____________________
4. April: D M D N O I A ____________________
5. May: D L A R E M E ____________________
6. June: L P R E A ____________________
7. July: Y B R U ____________________
8. August: P E R T O D I ____________________
9. September: E R I H P P A S ____________________
10. October: P L A O ____________________
11. November: W E L L Y O  Z A P O T ____________________
12. December: T U R E S O I Q U ____________________

# A Rocky Cycle

Rocks are always changing. Some of the changes are fast and some are slow. Match the words below to the numbered spaces in the rock cycle diagram. Then write the letter of the correct word in the space provided.

| | |
|---|---|
| a. compaction and cementation | e. melting |
| b. cooling | f. sediments |
| c. heat and pressure | g. weathering and erosion |
| d. magma | |

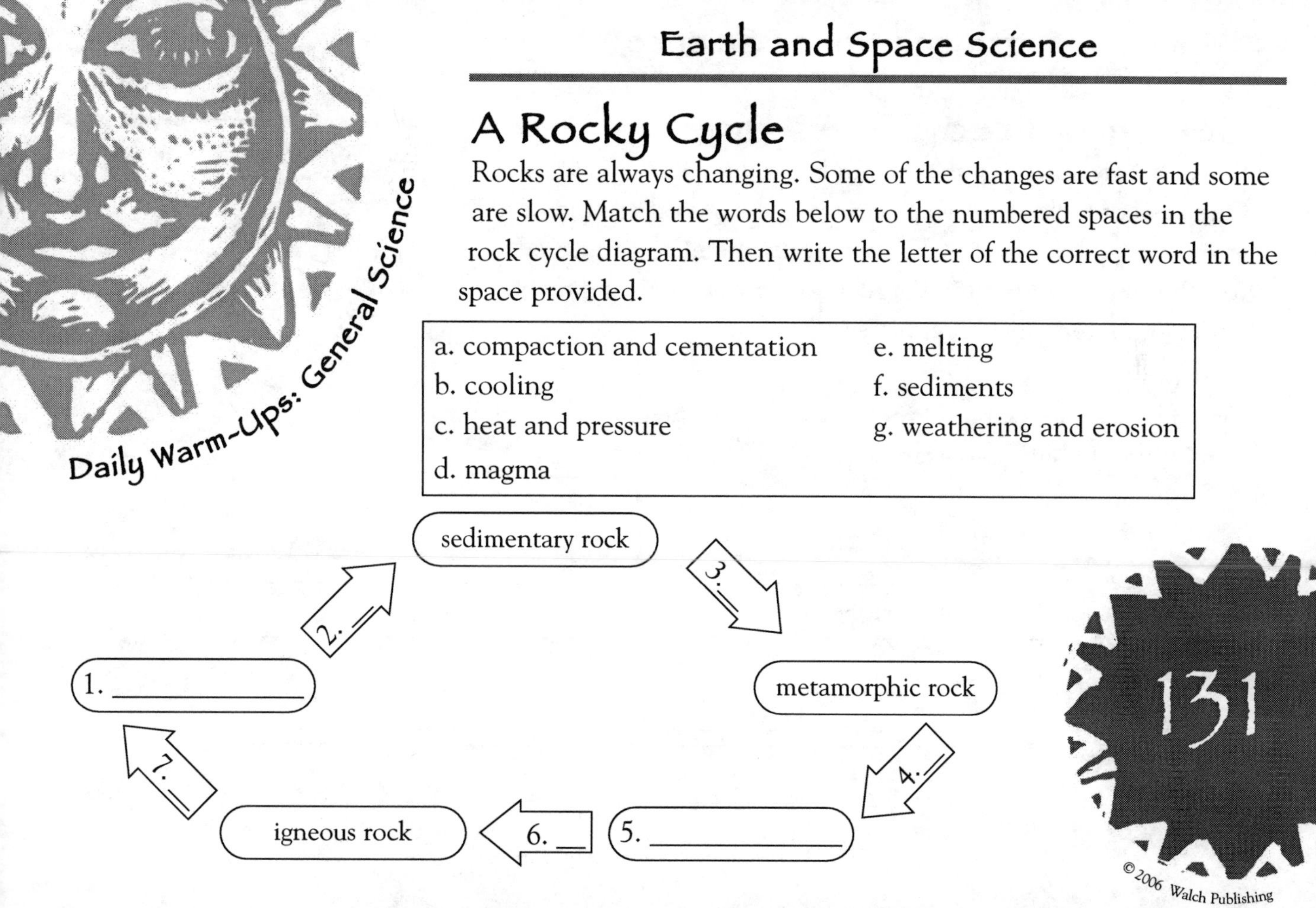

## Get Organized

Rocks are grouped into three main kinds, based on how they form. One kind of rock forms when magma cools. Another kind of rock forms when tiny bits of rock and other materials get squeezed together. Still another kind of rock forms when existing rock undergoes intense heat and pressure.

Organize the three kinds of rocks by filling in the concept map. Each kind of rock should have two examples.

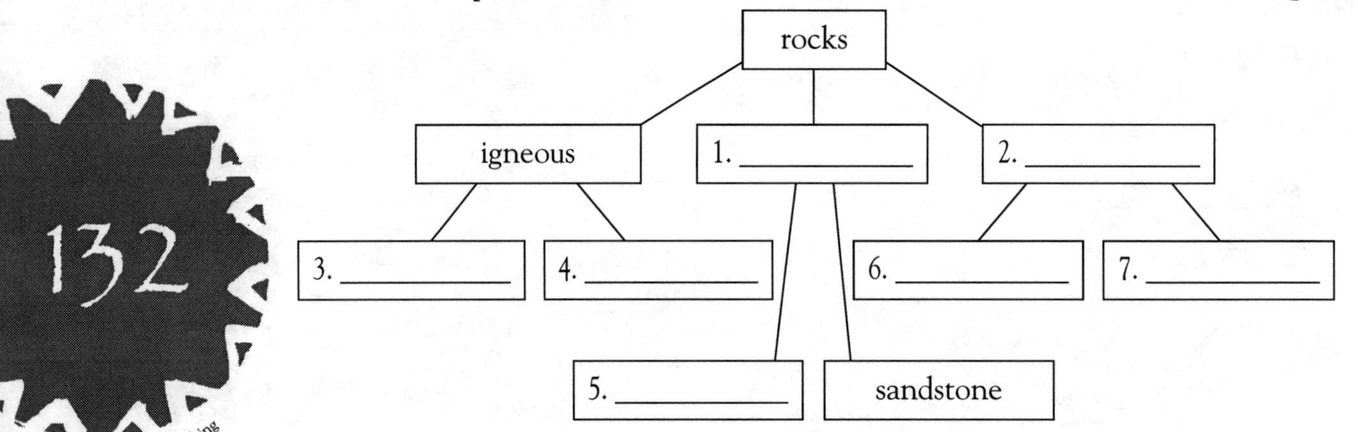

## Revealing Rock

There are two main kinds of igneous rock. One kind forms deep below Earth's surface. To reveal the name of these rocks, shade all the letters in the word *igneous* in the grid below.

After you have shaded in the grid, answer the following: What is the name of the kind of igneous rocks that form on Earth's surface? _______________ (*Hint:* Just change the first two letters of the word you spelled in the grid.)

| n | a | k | d | q | w | o | k | c | z | a | h | t | v | k | y | f | f | z | i | b | m | p | a | t | r | c | h |
|---|---|---|---|---|---|---|---|---|---|---|---|---|---|---|---|---|---|---|---|---|---|---|---|---|---|---|---|
| c | f | r | l | p | e | g | e | m | f | h | k | h | b | v | n | e | o | y | c | t | v | d | q | j | i | s | g |
| i | j | e | n | i | x | s | a | n | g | b | g | x | n | r | g | h | d | w | e | h | e | f | s | w | n | b | s |
| g | m | g | h | i | d | s | l | i | q | z | o | c | i | r | s | i | u | t | g | v | i | b | g | v | e | i | e |
| n | b | o | c | e | v | g | j | e | j | f | s | x | u | l | t | a | i | b | o | c | n | j | o | k | g | m | l |
| o | t | u | r | s | y | n | w | o | k | t | i | s | s | w | e | g | n | m | s | p | p | u | w | c | u | o | n |

Daily Warm-Ups: General Science

## I Never "Meta" Rock I Didn't Like

The word *metamorphic* comes from Greek words that mean "to change form." A *metamorphic rock* is a rock that has been changed, usually by heat and pressure. These changes turn the existing rock into a metamorphic rock.

Look at each rock named below. An arrow points to the name of the metamorphic rock that the rock can become. Some letters from the original rock are rearranged and included in the correct spaces. Fill in the other spaces to name each metamorphic rock.

1. shale → s l a _ e
2. granite → g n e i _ _
3. limestone → m _ _ _ l e
4. basalt → s _ _ _ _ t

## Analogies 1

It is often helpful to use analogies in science. An analogy is a comparison of something unfamiliar to something familiar. For example, you might say, "The earthquake shook the ground like a train passing by."

To understand the structure of the earth, it is helpful to compare it to a hard-boiled egg. Read each part of a hard-boiled egg below. Then write the part of Earth's structure that each part represents.

1. yellow yolk of the egg: ________________
2. white part of the egg: ________________
3. shell of the egg: ________________
4. cracked shell of the egg: ________________

# Analogies 2

Analogies can be helpful in understanding what happens at Earth's plate boundaries. Read each analogy below. Then write which plate boundary it refers to: divergent, convergent, or transform.

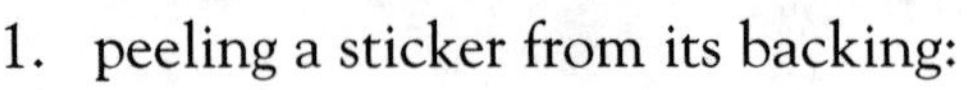

1. peeling a sticker from its backing: ________________

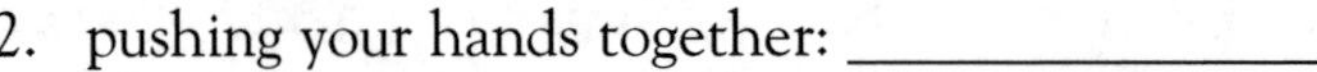

2. pushing your hands together: ________________
3. sliding your hands past each other: ________________
4. a rug crumpling when a door swings against it: ________________
5. a baseball player sliding into second base: ________________
6. crack formed when mud dries and pulls apart: ________________
7. car collision: ________________

## What's the Evidence?

The theory of plate tectonics explains how pieces of Earth's lithosphere move about on top of the mantle. There is much evidence that Earth's plates move.

Find the missing part of each incomplete word in the box and then write them on the lines provided.

| sils | ean | puz | rep |
|---|---|---|---|
| ther | joi | Ant | youn |

1. Some continents seem to fit together like pieces of a __ __ __ zle.
2. Fos __ __ __ __ of certain __ __ __ tiles were found on different continents, which could mean the continents were once __ __ __ ned.
3. Fossils of warm-wea __ __ __ __ plants were found in cold places such as __ __ __ arctica.
4. The __ __ __ __ gest rocks in the ocean are located at the mid-oc __ __ __ ridge.

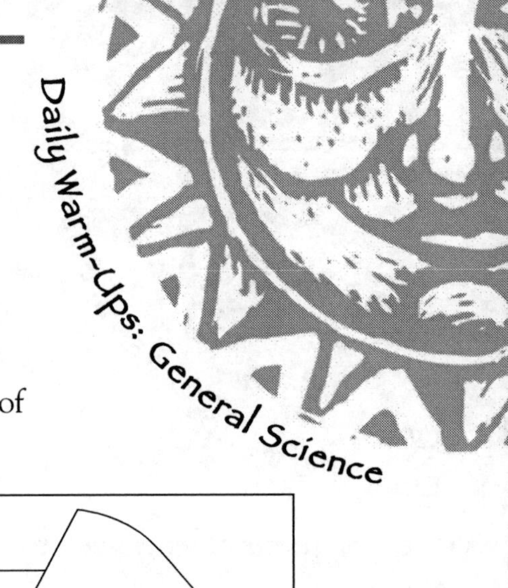


# Forced Up

Mountains are landforms that rise high above the surrounding land. Most mountains are formed by the forces associated with plate tectonics. Most of these forces occur near plate boundaries, but not all.

The diagrams show two kinds of mountains. Draw arrows on the diagrams to show the direction of the forces that cause each kind of mountain.

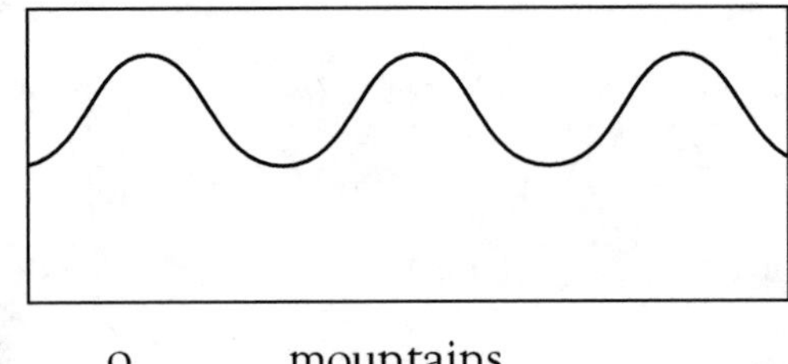

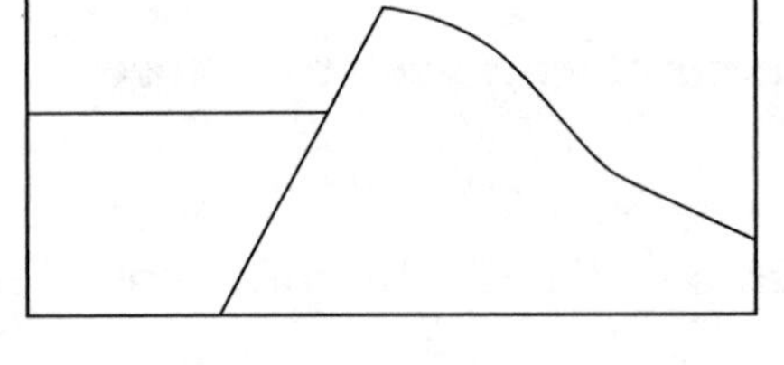

_ o _ _ _ mountains

_ a u _ _ - _ _ o _ _ mountains

Bonus: Fill in the consonants to name each kind of mountain.

## Pretty as a Picture

Forces build up mountains over millions of years. Eventually, those kinds of forces stop. But other forces continue to act on the mountains.

The drawing shows mountains that have been built up. They are tall and jagged. Draw how the mountains might look long after the constructive forces have stopped.

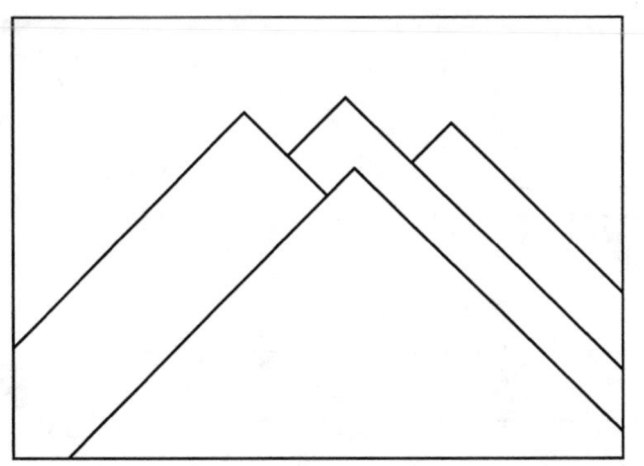

## Alike and Different—Earthquakes

An *earthquake* is the shaking of part of Earth's crust that results from the movement of rock beneath the surface. This movement usually occurs when huge slabs of rock move along a fault.

Each pair of words is related to earthquakes. Tell how the words in each pair are alike and how they are different.

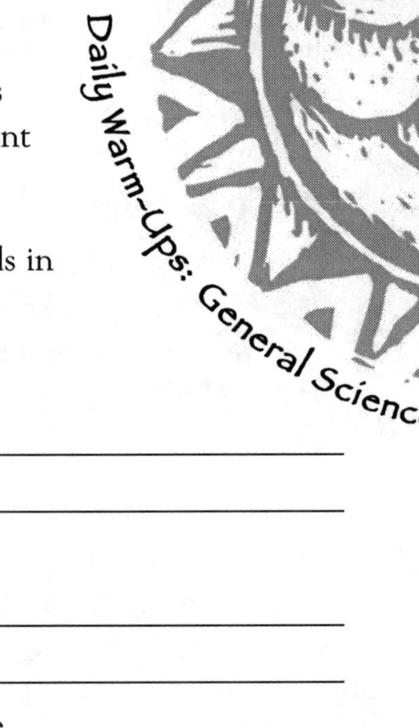


1. strike-slip fault : normal fault

   alike: ______________________________

   different: ______________________________

2. P waves : S waves

   alike: ______________________________

   different: ______________________________

3. Mercalli scale : moment magnitude scale

   alike: ______________________________

   different: ______________________________

# Earthquake Challenge

For each item in the game board below, you are given the answer and asked to provide the question. You may recognize this format, as it is similar to that of a popular television game show!

Try these questions (oops!—answers) below. Each one gets more difficult. See how many points you can get.

| **5 points**<br>An earthquake is caused when rocks suddenly move along this.<br>1. ______________________________ | **15 points**<br>This instrument detects earthquake waves.<br>3. ______________________________ |
|---|---|
| **10 points**<br>This is a large wave caused by an earthquake.<br>2. ______________________________ | **20 points**<br>These kind of seismic waves cannot travel through liquid.<br>4. ______________________________ |

## Location, Location, Location

How can scientists tell where an earthquake occurred? They use seismographs to compare the arrival times of P waves and S waves. Then they draw a circle on a map to show how far away the earthquake occurred. But it takes more than one circle.

The drawing shows three circles that scientists might draw on a map to locate an earthquake. Look at the map, and then answer the questions.

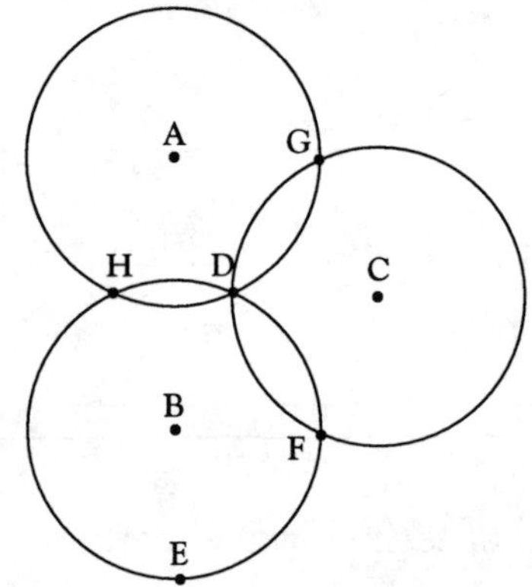

1. Which point is directly above where the earthquake started? _____
2. What is the name of this point? __________

# Volcanic Letter Trace

A *volcano* is an opening in the earth where molten rock comes to the surface. A volcano is also the mountain that builds up around this opening as the molten rock cools and hardens.

Use the clues to trace the correct word about volcanoes in each word box. The words can go in any direction, forward or backward. Then write the word in the space provided.

1.

| K | M | L | G |
|---|---|---|---|
| A | P | B | S |
| M | G | A | F |
| E | R | O | M |

2.

| G | O | F | F |
|---|---|---|---|
| N | S | H | I |
| I | T | W | R |
| R | D | J | E |

3.

| B | W | T | S |
|---|---|---|---|
| N | O | G | P |
| H | S | O | Q |
| U | T | F | R |

1. molten rock beneath the surface: ____________
2. belt of many volcanoes that forms a border around much of the Pacific Ocean: ____________
3. the kind of area in the mantle that formed the volcanoes of Hawaii: ____________

# Volcano Shapes

Different kinds of volcanic eruptions produce different kinds of volcanoes. The three kinds are *shield volcanoes*, *composite volcanoes*, and *cinder cones*.

Draw a simple diagram of each kind of volcano. Below the diagram, write the numbers of the phrases that describe that kind of volcano.

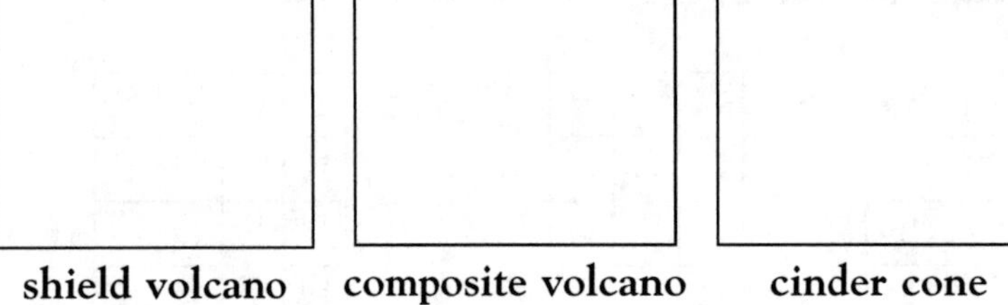

**shield volcano** ______ **composite volcano** ______ **cinder cone** ______

1. wide, gently sloping
2. small, steep-sided
3. tall, steep-sided
4. made of ash and cinders
5. eruptions of mostly basaltic lava
6. eruptions are most explosive

## The Most Active Volcano

Test your volcanic knowledge. Use the clues to fill in the words. Then use the letters in the circles to spell the most active volcano in the world. Write the letters in the order in which they appear, top to bottom and left to right.

1. magma that cuts across rock layers and hardens

   ___ ___ (◯) ___

2. volcano that is not expected to ever erupt again

   ___ ___ ___ (◯) ___ ___ ___

3. magma that reaches the surface

   (◯) (◯) ___ ___

4. volcano that erupted in Washington State in 1980

   ___ ___ (◯) ___ ___ S t . ___ ___ ___ ___ ___ ___

5. kind of energy provided by volcanic activity

   ___ (◯) ___ ___ ___ ___ ___ ___ (◯) ___

Most active volcano in the world: ____________________

# Mechanical or Chemical?

*Weathering* is a group of processes that break down rock on Earth's surface. The two types of weathering are mechanical and chemical. Mechanical weathering breaks rocks apart without changing their chemical makeup. Chemical weathering changes the chemical makeup of the minerals in the rock.

For each description of weathering below, write M if it is mechanical weathering and C if it is chemical weathering.

___ 1. breaking a rock with a hammer
___ 2. acid rain helping to dissolve rock
___ 3. decaying plants dissolving some minerals in rock
___ 4. repeated freezing and thawing of water in cracks of rocks
___ 5. rust stains on rock
___ 6. tree roots pushing against rock as tree grows
___ 7. animals digging small pits in rocks
___ 8. formation of caves

# The Correct Mixture

Soil is made of different-sized bits of weathered rock. Clay and silt are the smallest bits, and they pack together tightly. Sand grains are larger and pack more loosely. Soil also contains air, water, and organic matter, such as decayed leaves and animal wastes.

What's a good mixture of these components for growing most plants? The pie charts show the composition of two soils. Neither soil is good for growing most plants. Explain why beneath each graph. Then draw a third graph to show a better soil composition.

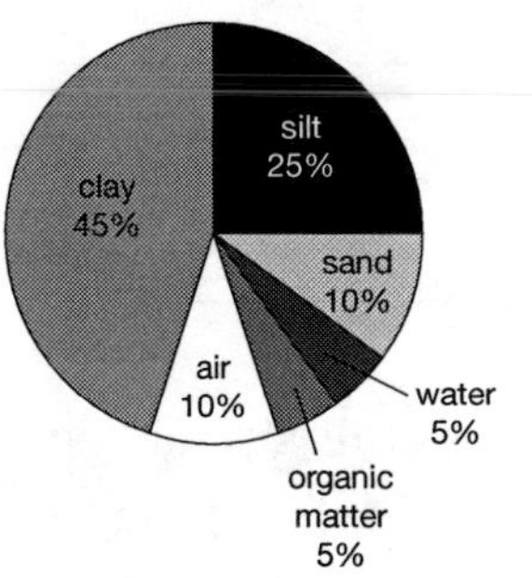


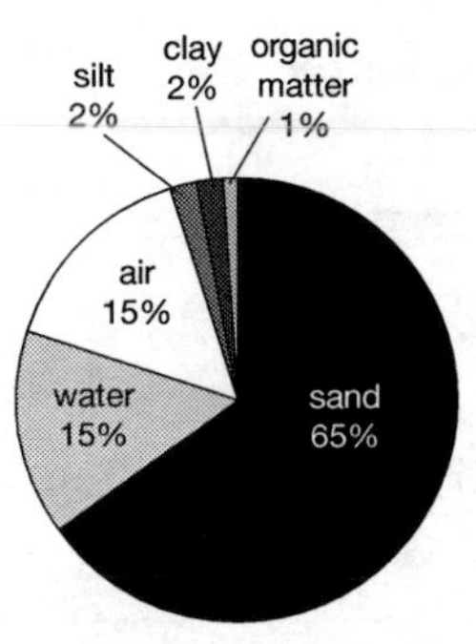


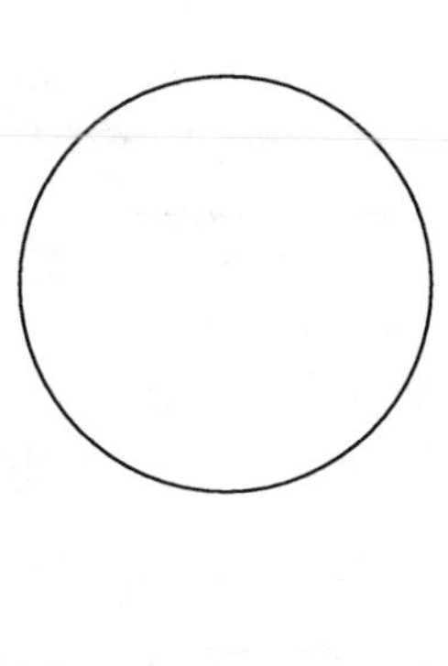

## Man's Best Friend

The dog has long been called "man's best friend." But that title is much more appropriate for the lowly earthworm. Why? Because earthworms help provide the fertile soil in which crops and other plants grow. When earthworms burrow in the soil, they loosen it and make openings through which air and water can enter. The worms also mix organic matter into the soil. Their wastes enrich the soil with minerals and chemicals that plants need. And when an earthworm dies, its body decays and further enriches the soil.

1. Why are earthworms so valuable? ___________________________

   ___________________________

   ___________________________

2. How might you use earthworms to help form a successful vegetable garden?

   ___________________________

   ___________________________

# May the Force Be With You

Weathering breaks down rocks into smaller particles. Then forces of erosion carry the particles away. The particles might be carried by wind, water, ice, or gravity. Eventually the particles come to rest. In this way, Earth's surface is constantly changing.

Circle the word in parentheses that makes each statement true.

1. The process in which particles settle out and come to rest is (erosion, deposition).
2. (Gravity, Friction) is the force that moves sediment downhill.
3. A fast movement of sediment and water downhill is a (mudflow, creep).
4. Valleys that have been carved out by a glacier usually have the shape of a (V, U).
5. The sediment from a river often creates a (meander, delta) where the river slows down and enters a large body of water.

## Nice Moves

No river is completely straight. It usually curves in a snakelike pattern. Each curve is called a *meander*. Generally, the flowing water erodes the outside of the meander and deposits sediment on the inside of the meander. This makes the meanders grow sideways and become more "curvy." Sometimes the outside of two meanders meet. The water then takes the straighter route and the meander gets cut off as an oxbow lake.



The drawing shows part of a river. Draw how it might have looked in the past.

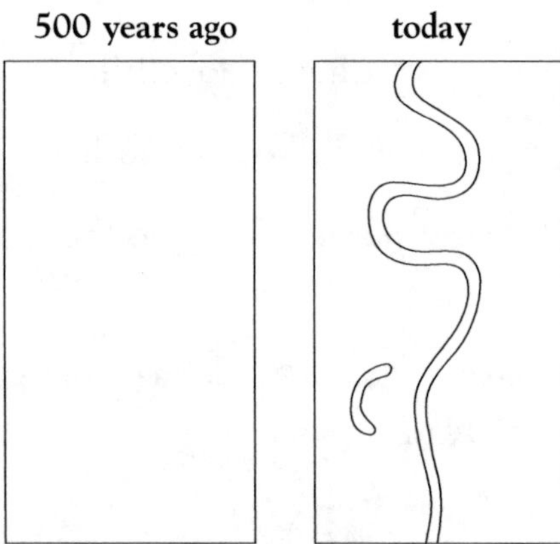

## More Nice Moves

The picture below shows how a certain river looks today. What will it look like 500 years from now? Keep in mind how a river erodes its banks and changes shape.

today

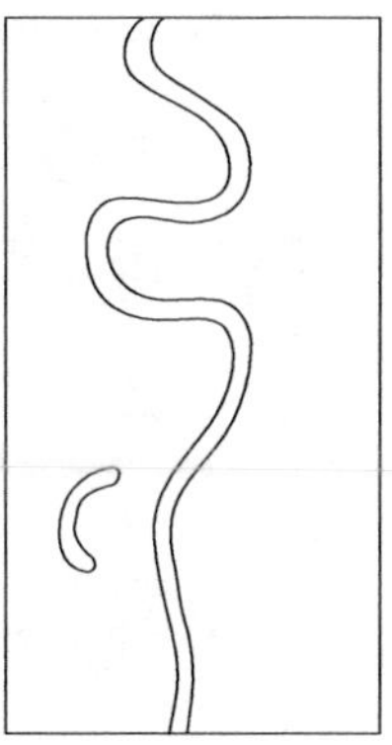

500 years from now

## Run River Run

One of the most important agents of erosion is running water—from small creeks that you can jump across to mighty rivers a mile wide. Running water shapes much of the landscape.

For each item below, write every other letter beginning with the first letter on line A. Break the letters into words that make sense. This is a clue. Use it to write a term about rivers on line B.

1. LBAENPDODWRBAMIZNHELDABIYKROIBVWEPRUSGYGSNTVEUM

   A. ______________________________

   B. ______________________________

2. DMERPKOISWIXTJALTRREIGVAEORPMTOSUSTEH

   A. ______________________________

   B. ______________________________

## Water Underground

When water falls as rain, some of it evaporates, some of it flows into rivers, and some of it soaks into the ground. The water gathers in the spaces between rocks and soil particles. This is groundwater. More than half of the United States gets their drinking water from groundwater.

Use the words in the box to label the diagram showing groundwater. Not all words will be used.

| aquifer | permeable layer | well |
|---|---|---|
| impermeable layer | water table | zone of saturation |

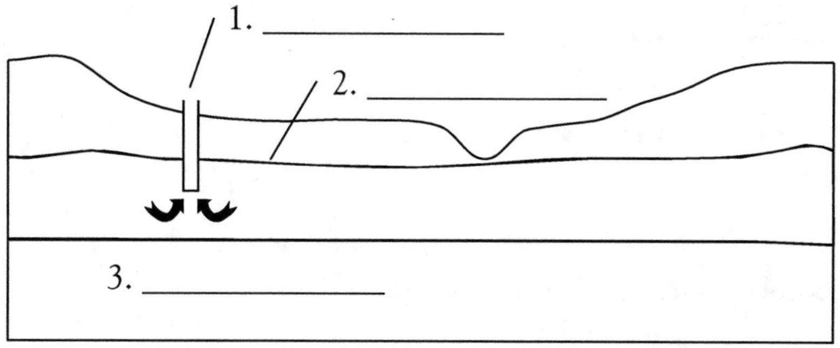

## Ice Is Nice

Among the most powerful agents of erosion are glaciers. A *glacier* is a mass of ice that moves slowly over the land. Glaciers shape the landscape by carving the land and leaving huge amounts of sediment when they melt.

See if you can find all the glacier words in this crossword puzzle.

**Across**

2. a sharpened peak left when glaciers carve the mountainsides
4. sediment deposited at the edge of a glacier, usually as a ridge
5. sediment of different-sized particles deposited by a glacier
6. type of glacier that occurs in mountains

**Down**

1. teardrop-shaped mound of glacial sediment smoothed as the glacier moved over it
3. lake formed as ice melts in a depression

## Think About the Word

Do you think science words are hard to learn? Some are. But many science words contain clues about what they mean. That makes them easier to remember. For example, *groundwater* means water that is in the ground.

The words below name features that are caused by wave erosion or wave deposition. Think about each word and what comes to mind when you read it. Then write what you think the definition is.

1. sea arch: ________________________________
2. sea cliff: ________________________________
3. headland: ________________________________
4. barrier island: ________________________________
5. spit: ________________________________

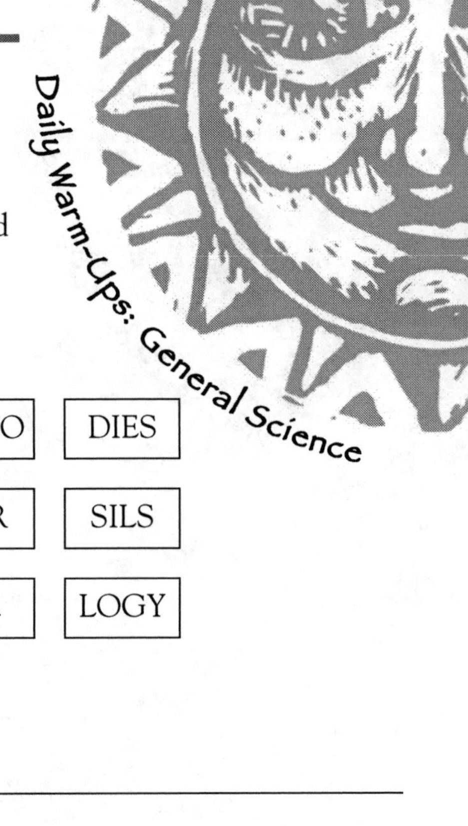


## Word Tiles Scramble

One of the many interesting branches of earth science is paleontology. *Paleontology* is . . . well, instead of telling you, have some fun figuring out the definition yourself. Unscramble the word tiles below, then write the definition on the lines.

| | | | | | |
|---|---|---|---|---|---|
| THAT | MS O | F LI | OF | ONTO | DIES |
| STU | ANCH | A BR | PALE | FOR | SILS |
| IS | NCE | EXT | INCT | FE . | LOGY |
| SCIE | FOS | AND | | | |

______________________________________________

______________________________________________

## Rocky Detective

When geologists look at layers of rock on a canyon wall, they often want to know how old the different layers are. Sometimes it's important just to know which layer formed first, second, and so on. If layers have not been overturned, the one on the bottom is the oldest.

Study the cross section below. The letters represent the formation of different rock and other events such as faults and erosion. Put the letters in order from the youngest to oldest event.

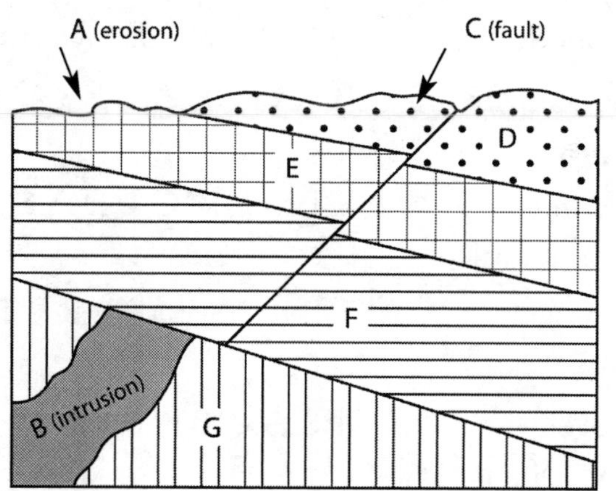


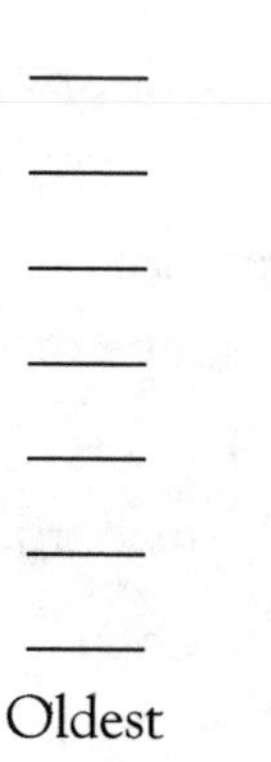


Youngest

____

____

____

____

____

____

____

Oldest

## Treasure in Rock

If you look closely at a sedimentary rock, such as limestone or shale, you may find fossils. *Fossils* are remains or traces of living things from the past.

Read the clues below about different kinds of fossils. Then circle the correct word in the string of letters that follows each clue. The word may read forward or backward.

| | |
|---|---|
| 1. means “turning to stone” | D E I F I R T E P U G H P A |
| 2. A dinosaur footprint is one of these kinds of fossils. | B A C T R A C E Y C S |
| 3. a hollow area in rock in the shape of the organism that died | R E D Y M O L D A S W B |
| 4. Insects are often preserved in this hardened tree sap. | Y E R P A M B E R S T A N D |
| 5. a copy of the shape of the organism; the opposite of a mold | T S A C B A S K E N |

Daily Warm-Ups: General Science

# How Can I Remember?

Geologists divide Earth's history into eras and periods of the geologic timetable. There are only four eras. But how can you remember all those periods?

Create a way to remember the periods in the correct order by thinking of a sentence whose words use the first letter of each period.

| Era | Period |
|---|---|
| Cenozoic | Quarternary |
| | Tertiary |
| Mesozoic | Cretaceous<br>Jurassic<br>Triassic |
| Paleozoic | Permian<br>Carboniferous<br>Devonian<br>Silurian<br>Ordovician<br>Cambrian |
| Precambrian | |

________________________________________

________________________________________

________________________________________

________________________________________

________________________________________

## Dinosaur Hunt

The Mesozoic era is called the "age of reptiles." The most dominant reptiles were the dinosaurs. Thousands of different kinds of dinosaurs existed. See if you can find eight of them in the word search below. Names can read forward, backward, up, or down. Look closely. One of the names turns a corner!

| | | | | | | | | | | | | |
|---|---|---|---|---|---|---|---|---|---|---|---|---|
| S | U | R | U | A | S | O | N | N | A | R | Y | T |
| R | M | A | I | A | S | A | U | R | U | S | E | R |
| E | T | L | O | G | C | N | I | B | R | U | B | I |
| X | D | L | Y | T | W | O | Q | C | A | R | J | C |
| P | I | O | N | H | U | R | A | Y | L | U | L | E |
| G | D | S | U | C | O | D | O | L | P | I | D | R |
| A | P | A | T | O | S | A | U | R | U | S | N | A |
| B | C | U | S | Q | A | B | T | L | I | P | X | T |
| M | P | R | O | D | I | G | A | S | C | I | R | O |
| E | N | U | L | R | U | A | S | O | R | E | T | P |
| C | I | S | T | E | G | O | S | A | U | R | U | S |

# Above and Beyond

Earth's atmosphere has four distinct layers. Think about in which layer you would find the following items. Then add each drawing to the correct layer in the diagram.

weather balloons

meteor trails ("shooting stars")

aurora borealis (northern lights) /////

the place you live

clouds and most weather

radio waves that bounce back to Earth

ozone

Thermosphere

Mesosphere

Stratosphere

Troposphere

## Over and Over Again

Water on Earth moves from the land to the oceans to the atmosphere in a never-ending water cycle. Label the water cycle diagram by adding arrowheads to the bands and using the words from the box.

| condensation | evaporation | precipitation | runoff |
|---|---|---|---|

# Weather Lingo

*Precipitation* is any form of water that falls from clouds to Earth's surface. There are several forms of precipitation. Discover three of them by playing Weather Lingo. Start with the given letter. Then add one letter at a time using the clues. You may have to rearrange the letters to get to the final word.

1. Start with: I

   not out: __ __

   famous dog: __ __ __

   (__ __ __ Tin Tin)

   name of precipitation: __ __ __ __

2. Start with: O

   negative: __ __

   immediately: __ __ __

   name of precipitation: __ __ __ __

3. Start with: T

   a wooden peg that holds a golf ball: __ __ __

   sounds like a place to sit, but no "a": __ __ __ __

   name of precipitation: __ __ __ __ __

## Choose Your Tool

What comes to mind when you think of weather? You might think about temperature and whether it's a sunny or a rainy day. Weather includes many different conditions. Each condition can be measured.

Unscramble the first part of each weather instrument below. Then match the instrument with what it measures by writing the letter in the space provided.

**Instrument**

___ 1. (Y G H O R) _ _ _ _ _ _ _ METER

___ 2. (M N O A E) _ _ _ _ _ _ _ METER

___ 3. (O M T R E H) _ _ _ _ _ _ _ _ METER

___ 4. (R O B A) _ _ _ _ _ METER

| a. wind speed | c. humidity |
|---|---|
| b. temperature | d. air pressure |

## Read the Map

How good are you at reading weather maps? It's easy if you know what the symbols mean and what kind of weather each symbol represents. Use the map below to answer the questions.

1. Which city is probably cooler, B or C? ____
2. Which two cities are the most likely to be experiencing storms? ______________
3. Which city is most likely having sunny weather? ____
4. Suppose city D has partly cloudy skies. What change in the weather would you predict for this city over the next couple days? ______________

______________________________

______________________________

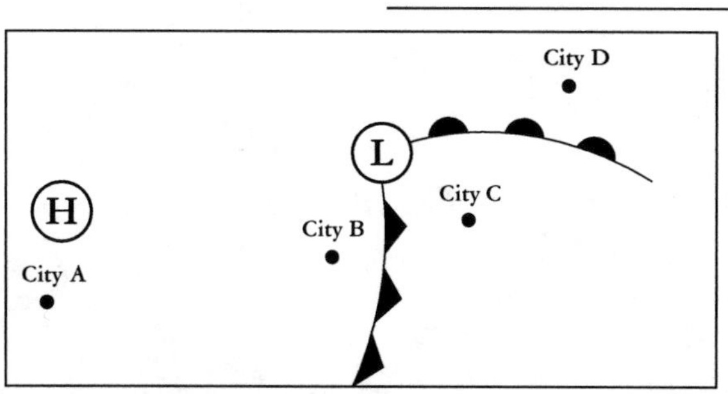


(H) High pressure
(L) Low pressure
Warm front
Cold front

Daily Warm-Ups: General Science

# It's Electrifying!

No matter where you live, you have probably been in a storm. Maybe it was a huge thunderstorm with brilliant flashes of lightning. Or maybe it was a frigid, blinding snowstorm.

Unscramble each word in the storm cloud. Then match each word with the correct clue below.

1. cloud that produces thunderstorms: _______________
2. small column of rotating air: _______________
3. begins over warm, tropical water: _______________
4. loud sound resulting from lightning: _______________
5. dome of water that sweeps over land during hurricane; a storm _______________

HRNTEDU
NRDTAOO
UGESR
RIUCNHEAR
SUMUCOILBMNU

# 'Tis the Season

Unless you live very close to the equator, your area has different weather from season to season. The temperature changes, the amount and kind of precipitation changes, and the amount of daylight changes.

Look at the diagram of Earth in its orbit. Then answer the following questions about seasons.

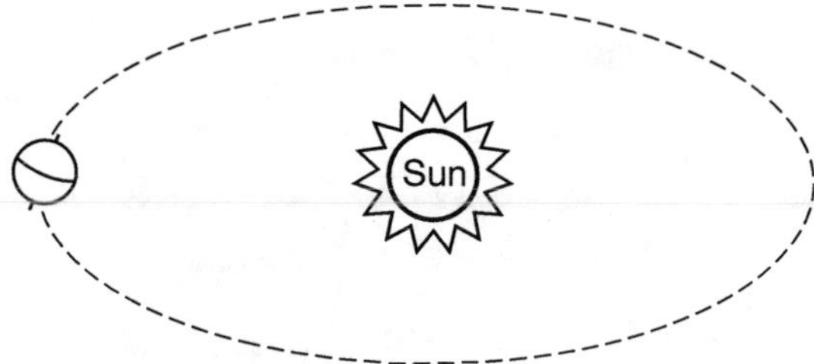


1. What season is it in the Northern Hemisphere? ____________
2. What are two things that cause seasons? ____________________________

______________________________________________

3. On the diagram, draw Earth to show it in the opposite season shown above.

## Ocean Motion

The ocean is in constant motion. You can see some of these motions just by standing on the shore. Other motions are harder to see, but they affect climates all around the world.

Use the clues to fill in the words about ocean motion. Then use the boxed letters to discover one of the forces that moves ocean water.

1. current of warm water in Atlantic: [ ]_ _ _   _ _ _ _ _ _ _
2. top of wave: _[ ]_ _ _
3. amount of salt in water: _[ ]_ _ _ _ _ _ _
4. carries energy through water: _ _[ ]_
5. daily rise and fall of ocean: _[ ]_ _
6. like a river in the ocean: _ _ _ _ _ _[ ]
7. the tendency of a body to float: _ _ _[ ]_ _ _ _

a force that moves water: _ _ _ _ _ _ _

## Deep Blue Sea

Less than a hundred years ago, people thought the ocean floor was flat. But technology has shown that the bottom of the sea has mountains, hills, canyons, and other features.

The diagram shows a simple profile of an ocean. Use the words in the box to label each feature. If the words are new to you, don't worry. Just think about what the word might mean and compare it to the different places shown in the diagram.

| abyssal plain | continental slope | seamount |
|---|---|---|
| continental shelf | mid-ocean ridge | trench |

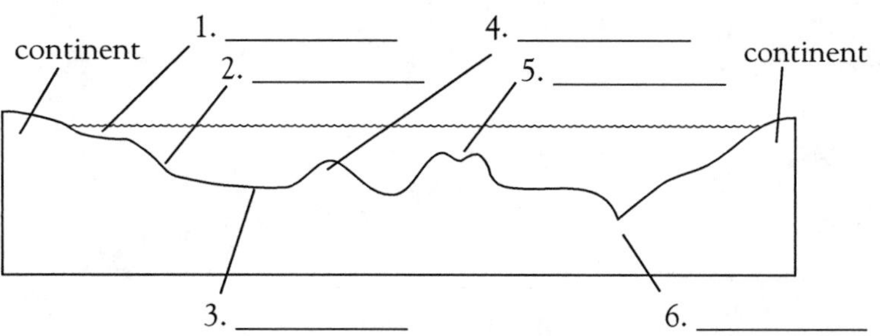

# The World Ocean

How many oceans are there? Did you say seven? No, that's the number of continents. Does five sound about right? That's one too many. There are four oceans. But because all the oceans connect, many scientists think of them as forming just one world ocean. The puzzle shows this idea. Fill in the connecting puzzle with the four ocean names.

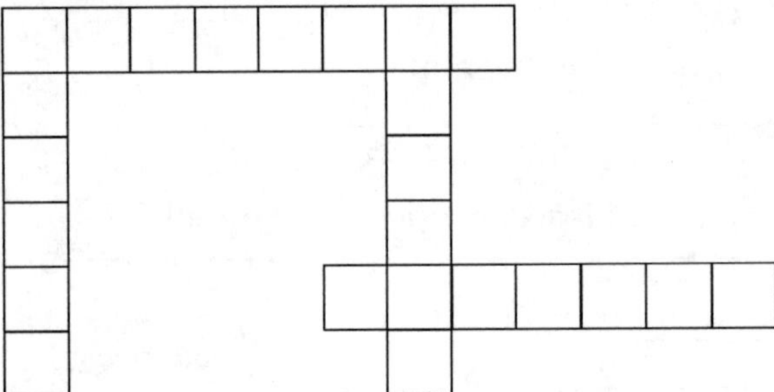

## What Makes It Curve?

If you look at a map of ocean currents, you notice that they travel in big circles. The currents move fairly straight east or west near the equator. But then they curve north or south along a continent. The currents curve because Earth rotates.

This effect of Earth's rotation on the direction of the currents has a name. Unscramble the letters below to find it. As a hint, the second and fourth letter of each word is in the correct position.

LOSICROI      CFEETF

_ _ _ _ _ _ _ _ _ _      _ _ _ _ _ _ _

## A Beautiful Word

The deep ocean is in complete darkness. Sunlight does not reach to the ocean depths. So how do deep-sea fish find food? Many of these organisms have body parts that produce their own light, just like a firefly. This special property has a long, fancy name. But like many long, fancy science words, each part of the word helps describe what's being named.



The word you're looking for uses part or all of the three words in the word equation below. The meanings in parentheses will help you figure out this beautiful 15-letter word.

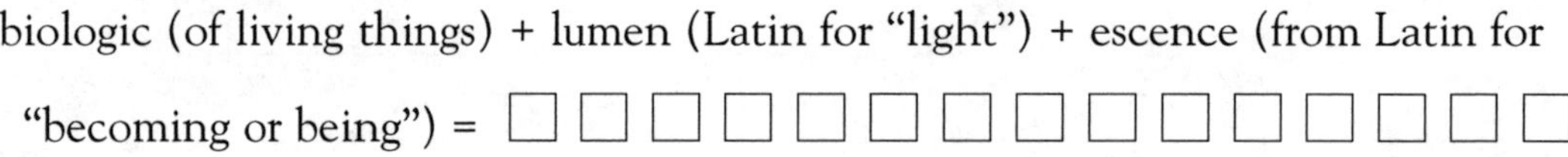

biologic (of living things) + lumen (Latin for "light") + escence (from Latin for "becoming or being") = □□□□□□□□□□□□□□□

# A Different Kind of Plant

What kind of a plant isn't green? A power plant! A power plant is a factory where electricity is produced from an energy source. The energy source might be uranium (nuclear energy), the sun (solar energy), falling water (hydroelectric energy), the wind, or fossil fuels. Most energy sources today are fossil fuels—oil, coal, and natural gas.

The steps below explain how a fossil fuel power plant works. But the steps are out of order. Put the steps in order by writing number 1 by the first step, 2 by the second step, and so on.

____ The spinning turbine turns a magnet in a generator.

____ The heat boils water to make steam.

____ The electricity flows out of the plant through power lines.

____ Coal, oil, or natural gas is burned to make heat.

____ The magnet spins near loops of wire and produces electricity.

____ The steam moves through a pipe and turns the blades of a turbine.

# A Thought Experiment

One of our most precious resources is soil. Susan did an experiment to figure out ways to reduce soil erosion. First, she spread out the same amount of soil in two trays. In one tray, she planted grass seeds. In a few weeks, the grass grew thick. Then she tilted both trays a little. She sprinkled the same amount of water on each tray. Soil started collecting at the bottom of one tray but not the other.

Think about Susan's experiment. Then answer these questions.

1. Which tray do you think had soil collecting at the bottom? ______________

2. Why didn't soil collect in the other tray?

   ______________________________________________

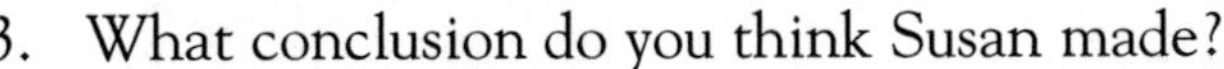

3. What conclusion do you think Susan made?

   ______________________________________________

4. How might she apply this conclusion to her gardening?

   ______________________________________________

## Dangerous Combinations

Some kinds of air pollution are caused by the combination of two or more "ingredients." The puzzle below includes some of these combinations. See if you can put the clues together to figure out the air pollution. Then unscramble the circled letters to name the law that helps reduce air pollution.

1. smoke + fog = _ _ _ _

2. hydrogen + carbon = _ _ _ (_) _ (_) (_) _ _ _ (_) _

3. hydrocarbons + nitrogen oxides + sunlight =

_ _ _ (_) _ _ _ (_) _ (_) _ (_) (_) _ _ _ _

4. sulfur dioxide + nitrogen oxides + moisture = (_) (_) _ _ _ _ _ _

Law: _ _ _ _ _ _ _ _ _ _ _ _

# Space Sudoku

A ______________________ is a person who travels in space.

To discover the missing word in the sentence above, unscramble the nine letters in the center of the sudoku puzzle below. Then fill in the puzzle so that each nine-by-nine square, each row, and each column use all the letters only once.

| | | | | | | | | |
|---|---|---|---|---|---|---|---|---|
| | A | | U | | | | | T |
| | U | | T | R | A | A | | |
| R | A | T | A | S | N | T | U | |
| O | | | S | T | A | | N | |
| T | N | A | A | T | R | | O | |
| | R | | N | U | O | T | | A |
| A | S | T | | A | T | N | | |
| | T | | R | | | | A | T |
| N | | | | A | | S | T | |

# Order in the Solar System

Our solar system includes the sun, eight planets, and other objects such as comets, meteors, and asteroids. The planets orbit the sun in a certain order. The planets are listed below alphabetically. Rewrite them so they are in the correct order from the sun.

**Planets**

Earth ______________

Jupiter ______________

Mars ______________

Mercury ______________

Neptune ______________

Saturn ______________

Uranus ______________

Venus ______________

## Star Challenge

For each item in the game board below, you are given the answer and are asked to provide the question. Try these answers below. Each one gets more difficult. See how many points you can get.

| **5 points**<br>A collection of billions of stars held together by gravity; example is the Milky Way<br>1. ____________ | **15 points**<br>An explosion of a supergiant star<br>3. ____________ |
|---|---|
| **10 points**<br>A pattern of stars in the sky; example is Ursa Major<br>2. ____________ | **20 points**<br>How bright a star looks to us in the sky<br>4. ____________ |

# Hey, You're In My Shadow

You know that the moon orbits Earth. But did you know that the moon's orbit is slightly tilted compared to Earth's orbit around the sun? That means the earth, sun, and moon usually do not directly line up. But every now and then, they do. Then we have eclipses.

The diagram below shows a solar eclipse. This happens when the moon is directly between the sun and Earth, casting a shadow on Earth. The other kind of eclipse is a lunar eclipse. Draw the positions of the earth, sun, and moon during a lunar eclipse.

**Solar Eclipse**

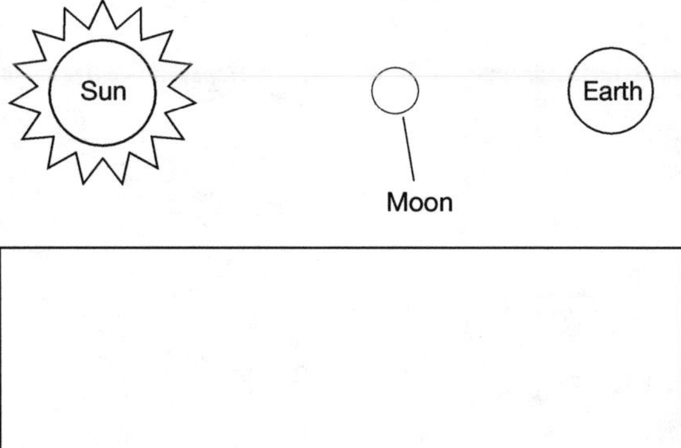


**Lunar Eclipse**

## Space Trace

Use the clues to trace the correct word about space in each word box. The words can go in any direction, forward or backward. Then write the word in the space provided.

1. telescope that uses mirrors to gather light: ____________

| T | A | M | W | A |
|---|---|---|---|---|
| G | F | E | C | T |
| E | U | L | O | I |
| R | E | F | D | N |
| P | X | H | L | G |

2. the movement of one object around another object: ____________

| F | I | N | R | B |
|---|---|---|---|---|
| L | O | V | E | B |
| U | Q | B | S | C |
| T | I | O | N | N |
| V | F | Y | V | K |

3. The Milky Way is this kind of galaxy: ____________

| G | E | J | Z | W |
|---|---|---|---|---|
| R | L | P | T | I |
| O | L | I | R | C |
| C | C | M | B | A |
| U | Z | E | J | L |

4. an explosion of gas from the sun's surface: ____________

| K | P | S | D | X |
|---|---|---|---|---|
| A | R | E | G | L |
| H | A | T | Q | I |
| F | L | E | D | Y |
| R | A | L | O | S |

1. 1. earth science
   2. physical science
   3. life science
   4. space science
   5. technology

   Number 3 should be circled.
2. 1. make a hypothesis
   2. draw conclusions
   3. identify the problem
   4. test the hypothesis
   5. gather information

   The order of statements is 3, 5, 1, 4, 2.
3. **SI Units**

| Unit | Symbol | Measures |
|---|---|---|
| gram | g | mass |
| meter | m | length |
| cubic meter | $m^3$ | volume |

**SI Prefixes**

| Prefix | Symbol | Means |
|---|---|---|
| milli- | m | 1/1000 |
| centi- | c | 1/100 |
| kilo- | k | 1000 |

4. 1. c
   2. d
   3. b
   4. a
   5. e
5. 1. c
   2. a
   3. d
   4. b
   5. g
   6. h
   7. f
   8. e
6. 1. safety goggles, heating, sharp
   2. glassware
   3. spills
   4. accident
   5. chemicals
   6. Bunsen burner, hot plate

7. 1. c
   2. d
   3. a
   4. e
   5. b
   6. f

8. 1. 1.86 m
   2. 8.48 qts
   3. 1.125 kg
   4. 25.07 yds
   5. 16.1 km
   6. 4 L milk

9. 1. meteorologist
   2. park ranger
   3. aeronautical engineer
   4. X-ray technician

10. 1. Geraniums grow best when watered once a week.
    2. the number of times watered

11. 1. d
    2. b
    3. a
    4. e
    5. c

    The word is *organism*.

12. a. cell membrane
    b. vacuole
    c. nucleus
    d. cytoplasm
    e. chloroplast
    f. cell wall

    The plant cell is cell 2.

13. 1. ribosomes
    2. mitochondria
    3. nuclear membrane
    4. vacuoles

14. 1. diffusion
    2. osmosis
    3. active transport

15.

| | |
|---|---|
| Rr | rr |
| Rr | rr |

Two offspring will produce round seeds.

16.

| SS | Ss |
|---|---|
| Ss | ss |

Chance of no disease: 75%
Chance of disease: 25%

17. The second circle should show the four paired chromosomes lined up in the middle of the cell. The nuclear membrane has disappeared. The third circle should show the paired chromosomes split and moved to opposite sides of the cell. The fourth circle should show a membrane around each group of single chromosomes. The cell is splitting or has split into two new cells.
Unscrambled term: nuclear membrane

18. First Stage order is: 4, 2, 5, 3, 1
Second Stage order is: 8, 7, 10, 6, 9

Four new cells are produced.

19. Sentences 2 and 4 complete the paragraph.

20.

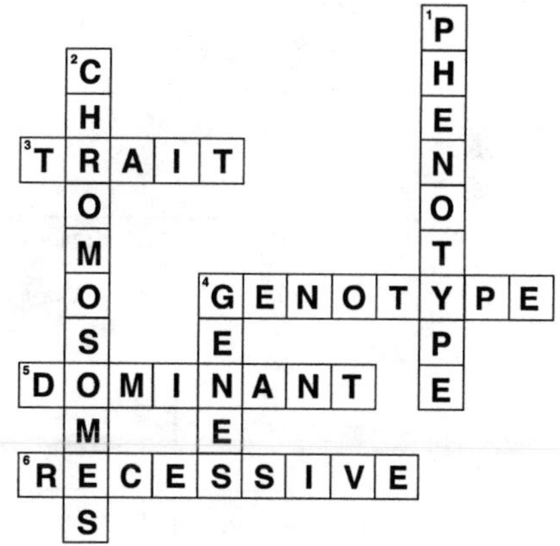


21. 1. descent
2. evolve
3. adaptation
4. fossils
5. natural
6. mutation

7. species
8. variation
9. Darwin

The name of the process is evolution.

22. 1. A
    2. H
    3. V
    4. H
    5. V

23.

| Kind of Evidence | Example | What Evidence Shows |
|---|---|---|
| fossil record | any series of related species that show gradual change over time, such as the development of the horse | Early species are different from recent species. |
| similar stages of development | Embryos of humans, chickens, and other vertebrates are similar. | All vertebrates descended from a common ancestor. |
| homologous structures | bones of human arm and bones of cat leg | Four-limbed animals with backbones shared a common ancestor. |

24. 1. c
    2. b
    3. d
    4. a

25.

<table>
<tr><td colspan="5">plants</td><td>animals</td></tr>
<tr><td colspan="3">protists</td><td colspan="2">plants</td><td>animals</td></tr>
<tr><td colspan="2">monerans</td><td>protists</td><td>plants</td><td>fungi</td><td>animals</td></tr>
<tr><td>eubacteria</td><td>archaebacteria</td><td>protists</td><td>plants</td><td>fungi</td><td>animals</td></tr>
</table>

26. A. all books
    B. travel books
    C. Japan guidebooks
    D. mysteries
    E. cookbooks

27. From bottom to top: kingdom, phylum, class, order, family, genus, species. Sentences will vary.

28. Answers will vary. Students might add apartment number as another level of classification. If they live in a rural area, they might not use city or street classification levels.

29. Possible categories for classifying the food items include frozen foods, meat, dairy, and produce.

30. 1. Alike: Both are microscopic.

Different: A virus is not a living thing; a bacterium is a living thing.

2. Alike: Both are kinds of bacteria or prokaryotes (single-celled organisms that lack a nucleus). Different: Archaebacteria are probably the ancestors of eubacteria.
3. Alike: Both are kinds of bacteria. Different: A bacillus is a rod-shaped bacterium, and a spirillum is a spiral-shaped bacterium.

31. Number of bacteria in the table should be 1, 2, 4, 8, 16, 32, 64, 128, 256, 512, 1024. The graph should be a sharply rising curve.

32. 1. Bacteria in plant roots change nitrogen into compounds the plants can use.
    2. Bacteria are used to produce foods such as cheese, yogurt, and sour cream.
    3. Bacteria help produce soil by decomposing dead organisms.
    4. Some bacteria digest oil and are used to clean up oil spills.

33. 1. c
    2. a
    3. e
    4. b
    5. d

34. 1. hyphae
    2. mushrooms
    3. mold
    4. mycelium

    Bonus question: penicillium

35.

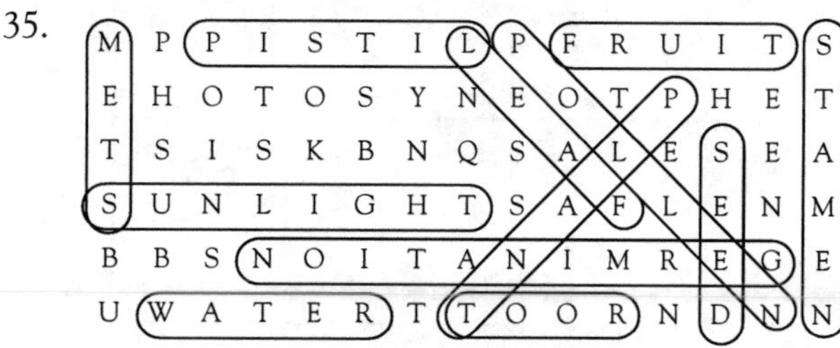


Plant process: photosynthesis

36.

| L | F | R | N | E | O | W | I | G |
|---|---|---|---|---|---|---|---|---|
| O | N | E | G | W | I | F | L | R |
| W | I | G | F | L | R | O | N | E |
| E | O | N | L | G | F | I | R | W |
| G | R | F | I | O | W | N | E | L |
| I | W | L | R | N | E | G | O | F |
| F | L | O | E | I | G | R | W | N |
| N | G | I | W | R | L | E | F | O |
| R | E | W | O | F | N | L | G | I |

37. stomata
38. 1. germination
    2. pollination
    3. fertilization
    Order is 2, 3, 1
39. 1. The thick stem stores water.
    2. Deep roots absorb water that sinks quickly through the sandy soil.
    3. By not carrying on transpiration, the spines don't release water.
40. tropism
41. stems: asparagus, celery; leaves: cabbage, lettuce, spinach; roots: carrots, radishes, sweet potatoes; seeds: corn, kidney beans, peanuts; flowers: broccoli, cauliflower; fruits: cherries, oranges, tomatoes
42. Riddle 1: e
    Riddle 2: a
    Riddle 3: b
43. metamorphosis
    complete metamorphosis
    incomplete (or gradual) metamorphosis
44.

| C | E | L | O | B | S | T | E | R |
|---|---|---|---|---|---|---|---|---|
| Y | L | F | R | E | T | T | U | B |
| L | A | G | F | Y | M | R | O | W |
| A | H | H | U | M | A | N | N | G |
| M | W | O | L | F | X | O | L | P |

Invertebrates: butterfly, worm, lobster
Vertebrates: human, whale, wolf

45. 1. Frog doesn't belong; reptile
    2. Whale doesn't belong; amphibian
    3. Penguin doesn't belong; mammal
    4. Rabbit doesn't belong; bird
    5. Shark doesn't belong; bony fish

46. Examples will vary.

| Mammal Group | Definition | Example |
|---|---|---|
| monotremes | 3 | platypus |
| marsupials | 1 | kangaroo |
| placentals | 2 | tiger |

47. 1. camouflage
    2. insects
    3. wind

48. Alike: Both are types of behavior.
    Different: Innate behavior is inherited and present at birth. Learned behavior is the result of experiences.

49. 1. conditioning
    2. insight
    3. Answers will vary. Possible answers: learning to ride a bike, tie shoes, hit a ball, or walk.

50. 1. endangered
    2. threatened

51. 1. tissue
    2. organ
    3. system
    4. Examples include blood, nerve tissue, and muscle tissue.
    5. Examples include heart, lungs, stomach, and brain.
    6. Examples include circulatory, skeletal, and respiratory systems.

52.

| System | Main Organs | Main Jobs |
|---|---|---|
| skeletal | bones | supports and protects the body |
| muscular | muscles | allows movement |
| digestive | mouth, esophagus, stomach, small intestine | breaks down food |
| circulatory | heart, blood vessels | sends blood throughout the body |
| respiratory | lungs | supplies the body with oxygen |
| nervous | brain, spinal cord, nerves | receives and sends messages |

53. 1. skull
2. backbone
3. ligaments
4. calcium
5. vertebrae
6. tendon
7. joint

54. 1. 4
2. 7
3. 1
4. 3
5. 8
6. 2
7. 6
8. 5

55. 1. heart
2. blood
3. vessels

56. 1. Sample sentence: I take in oxygen when I inhale and get rid of carbon dioxide when I exhale.
2. Sample sentence: Air enters and leaves the body through the nose and mouth.
3. Sample sentence: Bronchi are small passageways that carry air into the lungs.

57. brain, spinal cord

58. 1. iris
2. pupil
3. cornea
4. lens
5. retina
6. optic nerve

59. 1. 4
2. 2
3. 1
4. 5
5. 3

60. 1. gallbladder
    2. diaphragm
    3. liver
    4. pancreas
    5. ventricles
    6. sperm
    7. hemoglobin

61.

| R | C | A | R | B | O | H | Y | D | R | A | T | E | S |
|---|---|---|---|---|---|---|---|---|---|---|---|---|---|
| E | Z | Y | S | U | V | E | N | D | E | R | B | A | X |
| T | A | P | R | O | T | E | I | N | S | I | Q | J | Y |
| A | B | S | S | N | I | M | A | T | I | V | O | V | F |
| W | O | U | L | P | M | J | G | C | X | A | N | M | L |
| S | T | A | F | T | M | I | N | E | R | A | L | S | Y |

62. 1. b
    2. i
    3. d
    4. g
    5. e
    6. c
    7. f
    8. a
    9. h

63. 1. b
    2. e
    3. d
    4. f
    5. c
    6. a

64. 1. Get eight hours of sleep every night.
    2. Cover your mouth when you sneeze or cough.
    3. Keep your hands away from your eyes, nose, and mouth.
    4. Eat a well-balanced diet with a variety of foods.
    5. Get regular exercise.
    6. Wash your hands after using the bathroom and before eating.

65. The arrows should start with the sun and lead to the wheat plant, the mouse, the snake, and the hawk in that order.

66. Food webs will vary. One possible food web would show the following: Arrows going from the grass to the deer, rabbit, and grasshopper. Arrows going from the rabbit, grasshopper, and frog to the snake. Arrows going from the deer, rabbit, snake, and owl to the coyote. Arrows going from the rabbit, grasshopper, frog, and snake to the owl.

67. 1. A. habitat; B. niche
    2. A. producer; B. consumer

68. 1. boreal
    2. aquatic
    3. grassland
    4. tropical
    5. desert
    6. tundra

    Mystery Word: biosphere

69. 1. herbivore
    2. omnivore
    3. carnivore

70. 1. about 500 million people
    2. The population increased sharply.
    3. Answers will vary but may include the ideas that industry and new inventions made life easier and increased food production, and advances in medicine allowed people to live longer.
    4. about 6.5 billion people

71. Circled words are property, state, density, and molecule.

72. 1. The mass of the apple is 235 g.
    2. The rock should be balanced with masses of 100 g, 25 g, 10 g, and 10 g.

73. A. 5 mL
    B. 32 mL
    C. Cylinder should be shaded to show a volume of 9.5 mL.
    D. Cylinder should be shaded to show a volume of 77 mL.

74. 1. 11.3 g/cm$^3$
    2. 11.3 g/cm$^3$
    3. The pieces of metal are the same metal because they have the same density.

75. The solid should contain circles that are close together in a regular pattern. There may be lines that indicate the particles vibrate in place. The liquid should show the particles farther apart and moving past one another, perhaps with arrows. The gas should show particles even farther apart and moving farther.

76. 1. nucleus
    2. electron
    3. neutron
    4. proton

77. 1. 6
    2. 6
    3. 6
    4. 12

78. Both drawings should have one proton and one electron. Additionally, the first drawing should have one neutron, and the second drawing should have two neutrons.

79. 1. lead, Pb
    2. gold, Au
    3. sodium, Na
    4. silver, Ag
    5. tungsten, W
    6. iron, Fe

80. Statements 1, 3, and 4 are true. The element symbol is H, for hydrogen.

81.

| Compound | Formula | Number of Atoms |
|---|---|---|
| water | $H_2O$ | 2 hydrogen, 1 oxygen |
| salt | NaCl | 1 sodium, 1 chlorine |
| chalk | $CaCO_3$ | 1 calcium, 1 carbon, 3 oxygen |
| sugar | $C_{12}H_{22}O_{11}$ | 12 carbon, 22 hydrogen, 11 oxygen |
| baking soda | $NaHCO_3$ | 1 sodium, 1 hydrogen, 1 carbon, 3 oxygen |

82. 1. Alike: They are both combinations of matter.
    Different: Matter in a compound is joined chemically; substances in a mixture are not joined chemically.
    2. Alike: They are both mixtures.
    Different: In a solution, particles in one substance are dissolved in another; in a suspension, the particles are not dissolved.
    3. Alike: They are both parts of a solution.
    Different: The solvent is the substance that dissolves the solute.

83. 1. Cu and Cl (copper and chlorine)
    2. $CuCl_2$ (copper chloride)
    3. produces or yields
    4. 1
    5. 2

84. 1. balanced
    2. $2H_2 + O_2 \rightarrow 2H_2O$
    3. $C_6H_{12}O_6 \rightarrow 6C + 6H_2O$

85. 1. exothermic
    2. endothermic
    3. single-replacement
    4. double-replacement

86.

| | | | | | | | | |
|---|---|---|---|---|---|---|---|---|
| T | R | L | U | P | E | H | N | A |
| P | H | N | T | A | L | E | R | U |
| E | A | U | N | H | R | P | L | T |
| L | N | E | P | T | U | A | H | R |
| A | P | T | R | N | H | U | E | L |
| H | U | R | L | E | A | T | P | N |
| R | L | H | E | U | T | N | A | P |
| U | E | P | A | L | N | R | T | H |
| N | T | A | H | R | P | L | U | E |

87. 1. B
2. B
3. A
4. A
5. B
6. A
7. A
8. B
9. A
10. B

88. 1. 2 m/s
2. 50 mi/h
3. 0.23 mi/minute; 14 mi/h

89. Car A: 50 mi/hr; Car B: 45 mi/hr; Car C: 30 mi/hr

90. 1. A
2. C
3. C
4. A
5. A. Drawing should show a large arrow pointing to the right. B. Drawing should show an arrow about the size of the smaller arrow pointing to the left. C. Net force is zero.

91. 1. 3
2. 1
3. 3
4. 2
5. 2
6. 3
7. 1
8. 1

92. satellite

| | | | | | | | | |
|---|---|---|---|---|---|---|---|---|
| I | L | E | S | T | E | A | L | T |
| E | T | T | A | L | I | S | L | E |
| S | L | A | T | L | E | T | E | I |
| T | T | L | E | A | L | I | E | S |
| E | E | I | L | S | T | L | A | T |
| L | A | S | I | E | T | L | T | E |
| T | S | T | L | E | L | E | I | A |
| A | I | L | E | T | S | E | T | L |
| L | E | E | T | I | A | T | S | L |

93. 1. The balls will hit the ground at the same time.
    2. The baseball and tennis ball will hit the ground at the same time. The sheet of paper will hit the ground later.
    3. air resistance
    4. mass, distance
    5. Sir Isaac Newton

94. work equals force times distance (or work = force × distance)

95. input, output, amount, direction, distance

96.

| | | |
|---|---|---|
| A 2 | B 9 | C 4 |
| D 7 | E 5 | F 3 |
| G 6 | H 1 | I 8 |

Magic Number: 15

97. Class 1: crowbar, scissors, seesaw
    Class 2: bottle opener, paper cutter, wheelbarrow
    Class 3: baseball bat, broom, rake

98. A. MA = 1
    B. MA = 2
    C. MA = 2
    D. MA = 3

99. 1. wedge
    2. inclined plane

3. pulley
4. screw
5. first-class, third-class

100.

| Compound Machine | Simple Machines |
|---|---|
| scissors | wedges, levers |
| fork | wedges, lever |
| hand-held can opener | wedge, lever, wheel and axle |
| bicycle | levers, screws, wheels and axles |

101. 1. kinetic
2–6. The following forms of energy can be placed in any order in this section of the concept map: chemical, heat, light, nuclear, electrical.

102. mechanical: wind, kicking a ball; heat: fire, the sun, boiling water; light: fire, lightning, the sun; chemical: gasoline, battery, head of a match; electrical: lightning, moving electrons; nuclear: the sun

103. 1. electrical, heat
2. chemical, mechanical
3. electrical, mechanical

104. The Law of Conservation of Energy

105. 1. You both did the same amount of work because you both used the same amount of force to move the leaves over the same distance.
2. You used more power because power is the rate at which work is done. You did the work in less time so your rate of work is higher.

106. 1. radiation
2. conduction
3. convection
4. Celsius
Mystery Word: calorie

107. insulator

| O | R | N | S | A | I | T | L | U |
|---|---|---|---|---|---|---|---|---|
| U | T | A | R | L | N | I | O | S |
| L | I | S | T | U | O | N | R | A |
| A | L | T | U | I | S | R | N | O |
| N | U | R | A | O | L | S | T | I |
| I | S | O | N | T | R | U | A | L |
| T | O | U | I | R | A | L | S | N |
| R | N | L | O | S | U | A | I | T |
| S | A | I | L | N | T | O | U | R |

108. Procedures will vary. Possible procedure: Put the same amount of hot water in each cup and place each cup inside a coffee can. Stuff a different insulating material between the outside of the cup and the inside of the can. Include one can with just air as an insulator. Place a thermometer in each cup. Record the temperature every two minutes and see which cup of water retains its heat the longest.

109.

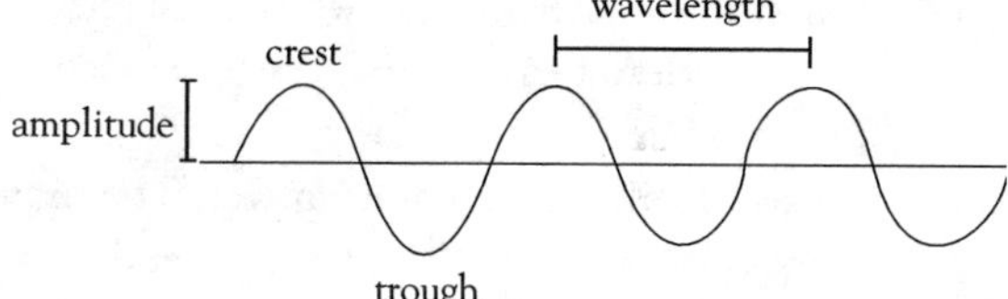


110.

111. The colors of the visible spectrum should be red, orange, yellow, green, blue, indigo (dark violet-blue), and violet.
112. Order from quiet to loud: whisper, library, normal conversation, heavy traffic, car horn, thunder overhead, jet engine nearby
113. Answers will vary. Sample answer: The drawing shows how sonar works. The ship sends out sound waves. The sound waves reflect off the ocean floor and return to a receiver on the ship. By knowing how fast sound travels in water, and how long it took the sound waves to return, you can tell how deep the water is at that point.
114. The first drawing should show more negatively charged than positively charged particles on the shoe and body. The carpet should have more positively charged particles because it gave up some of its negative particles to the shoe. The second drawing should show more negatively charged than positively charged particles on the hand. Negative particles are in the air between the hand and doorknob, attracted to the positive particles on the doorknob nearest the hand.
115. The two circuits are series circuit and parallel circuit. The series circuit should show wires connecting the three bulbs one after another with the first bulb connected to one terminal of the battery and the last bulb connected to the other terminal. The parallel circuit should show that each bulb has its own path to the battery.
116. 1. watt 3. open
    2. resistance 4. transformer
117. 1. July reading: 6124 kWh; August reading: 7510 kWh
    2. 1386 kWh
    3. $207.90
118. Across: cobalt, iron, magnetite, steel, nickel; Down: rubber, glass, gold, plastic, wood
119. 1. generator
    2. compass
    3. motor

120. 1. e
2. c
3. a
4. b
5. d

121. 1. compass
2. south
3. equator
4. east
5. latitude
atlas

122. 1. true
2. true
3. false
4. false
5. false

123. The clocks should show the following times: Eastern: noon; Central: 11 A.M.; Mountain: 10 A.M.; Pacific: 9 A.M.
1. Earth rotates from west to east.
2. 24

124. 1. hill
2. 20 feet
3. 884 feet
4. The east side because it is the steepest, as shown by the closely drawn contour lines.

125. The profile should start at an upper elevation at point A, then descend and rise again at a steeper slope to a higher elevation at point B.

126. 1. solids
2. naturally
3. chemical
4. alive
5. atoms

127. 1. Both are solids.
2. Both are not made of living things.

128. 1. quartz
2. feldspar
3. mica
4. calcite
5. dolomite
6. halite
7. gypsum
8. olivine

129. From hardest to softest, the minerals are C, D, A, B.

130. 1. garnet
2. amethyst
3. aquamarine
4. diamond
5. emerald
6. pearl
7. ruby
8. peridot
9. sapphire
10. opal
11. yellow topaz
12. turquoise

131. 1. f
2. a
3. c
4. e
5. d
6. b
7. g

132. 1. sedimentary
2. metamorphic
3–4. Examples might include granite, basalt, gabbro, and obsidian.
5. Examples might include shale, limestone, and conglomerate.
6–7. Examples might include slate, marble, gneiss, and quartzite.

133. Igneous rocks that form on Earth's surface are extrusive rocks.

| n | a | k | d | q | w | o | k | c | z | a | h | t | v | k | y | f | f | z | i | b | m | p | a | t | r | c | h |
|---|---|---|---|---|---|---|---|---|---|---|---|---|---|---|---|---|---|---|---|---|---|---|---|---|---|---|---|
| c | f | r | l | p | e | g | e | m | f | h | k | h | b | v | n | e | o | y | c | t | v | d | q | j | i | s | g |
| i | j | e | n | i | x | s | a | n | g | b | g | x | n | r | g | h | d | w | e | h | e | f | s | w | n | b | s |
| g | m | g | h | i | d | s | l | i | q | z | o | c | i | r | s | i | u | t | g | v | i | b | g | v | e | i | e |
| n | b | o | c | e | v | g | j | e | j | f | s | x | u | l | t | a | i | b | o | c | n | j | o | k | g | m | l |
| o | t | u | r | s | y | n | w | o | k | t | i | s | s | w | e | g | n | m | s | p | p | u | w | c | u | o | n |

134. 1. shale → slate
 2. granite → gneiss
 3. limestone → marble
 4. basalt → schist

135. 1. core
 2. mantle
 3. crust
 4. tectonic plates

136. 1. divergent
 2. convergent
 3. transform
 4. convergent
 5. transform
 6. divergent
 7. convergent

137. 1. puzzle
 2. Fossils, reptiles, joined
 3. weather, Antarctica
 4. youngest, ocean

138.

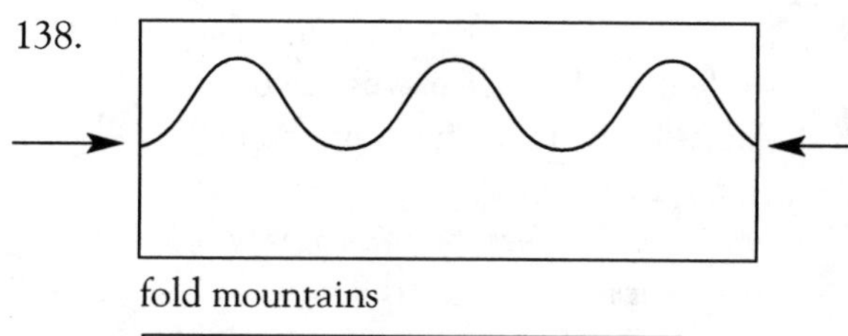

fold mountains

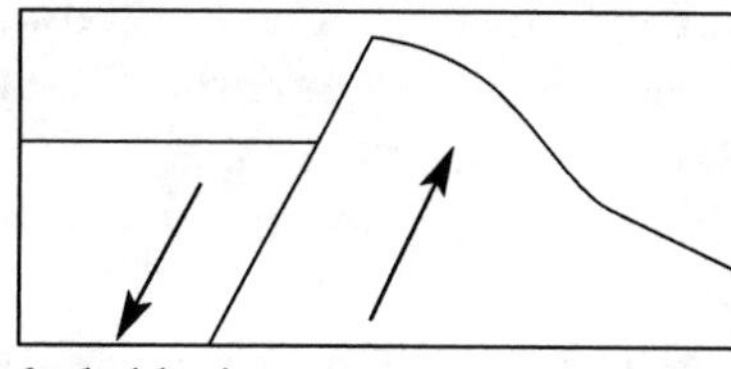

fault-block mountains

139. Drawings should show the same mountains but shorter and rounded, as a result of the forces of erosion.

140. 1. Alike: Both are faults that rock moves along to cause earthquakes.
 Different: In a strike-slip fault, rocks on either side of the fault move sideways. In a normal

fault, the rock on one side of the fault moves up while the other side moves down.

2. Alike: Both are seismic waves caused by an earthquake.
   Different: P waves, or primary waves, are the first waves to reach Earth's surface. They compress and expand. S waves, or secondary waves, reach the surface later than P waves. They move from side to side.
3. Alike: Both are scales used to measure an earthquake.
   Different: The Mercalli scale measures the strength of the earthquake at a certain place. The moment magnitude scale measures the total energy released by the earthquake. It is used most today.

141. 1. What is a fault?
2. What is a tsunami?
3. What is a seismograph?
4. What are S waves?

142. 1. D
2. epicenter

143. 1. magma
2. Ring of Fire
3. hot spot

144. Drawings should show a shield volcano to be broad and gently sloping, a composite volcano to be tall and steep, and a cinder cone to be small and steep.
shield volcano: 1, 5
composite volcano: 3, 6
cinder cone: 2, 4

145. 1. dike
2. extinct
3. lava
4. Mount St. Helens
5. geothermal
Most active volcano: Kilauea

146. 1. M 5. C
2. C 6. M
3. C 7. M
4. M 8. C

147. The large amount of silt and clay in the first soil wouldn't allow water to drain enough. The large amount of sand in the second soil wouldn't hold water well enough. Students' drawings should show a better balance between silt, clay, and sand, between 10 and 20 percent for each. It should also show at least 5 percent organic matter and 20 percent each of air and water.

148. 1. Earthworms improve and enrich soil by loosening it, mixing it, and adding organic matter to it.
    2. You could add earthworms to the soil to loosen it and enrich it.

149. 1. deposition
    2. Gravity
    3. mudflow
    4. U
    5. delta

150. Drawings might show the oxbow lake as part of a meander. The other meanders might not be as curvy as today.

151. One of the meanders might be cut off to form an oxbow lake. Drawings should show differences in the position and amount of curve in the meanders.

152. 1. A. deposit at river mouth; B. delta
    2. A. land drained by river system; B. drainage basin

153. 1. well
    2. water table
    3. impermeable layer

154.

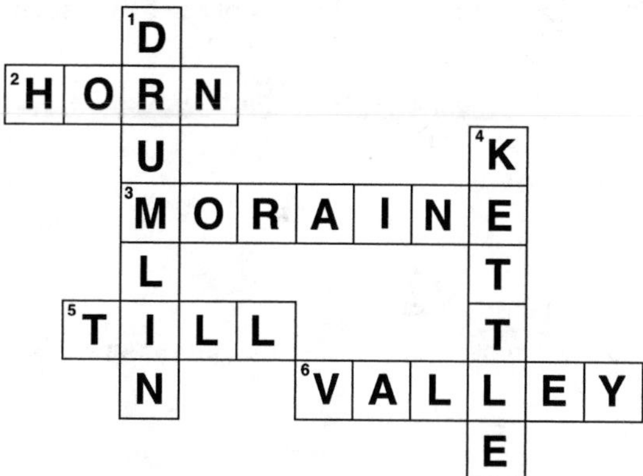

155. 1. a feature created when waves erode a rocky cliff into the shape of an arch
 2. a cliff formed by erosion of waves along the shore
 3. a part of the shore that sticks out into the ocean
 4. an island that runs parallel to the shore and forms a barrier against storms
 5. a sandy deposit that sticks out into the ocean like a finger

156. Paleontology is a branch of science that studies fossils and extinct forms of life.

157. Youngest to oldest: A, C, D, E, F, B, G

158. 1. petrified
 2. trace
 3. mold
 4. amber
 5. cast

159. Accept all reasonable mnemonic devices.

160.

```
S U R U A S O N N A R Y T
R M A I A S A U R U S E R
E T L O G C N I B R U B I
X D L Y T W O Q C A R J C
P I O N H U R A Y L U L E
G D S U C O D O L P I D R
A P A T O S A U R U S N A
B C U S Q A B T L I P X T
M P R O D I G A S C I R O
E N U L R U A S O R E T P
C I S T E G O S A U R U S
```

161. Troposphere: the place you live, clouds and most weather; Stratosphere: weather balloons, ozone; Mesosphere: meteor trails; Thermosphere: aurora borealis, radio waves that bounce back to Earth

162.

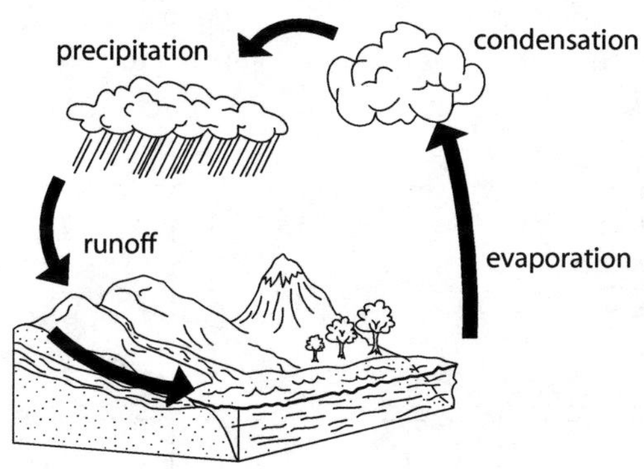


163. 1. IN, RIN, RAIN
2. NO, NOW, SNOW
3. TEE, SEET, SLEET

164. 1. hygrometer, c
2. anemometer, a
3. thermometer, b
4. barometer, d

165. 1. B
2. B and C
3. A
4. The weather would likely get cloudier, with possible precipitation, as the warm front and then the cold front passes. The temperature might get a little warmer, then cooler.

166. 1. cumulonimbus
2. tornado
3. hurricane
4. thunder
5. surge

167. 1. summer
2. the tilt of Earth and its orbit around the sun
3. The drawing should show Earth on the opposite side of the sun, but the tilt of the axis should be the same.

168. 1. gulf stream
2. crest
3. salinity
4. wave
5. tide
6. current
7. buoyancy

boxed word: gravity

169. 1. continental shelf
2. continental slope
3. abyssal plain
4. seamount
5. mid-ocean ridge
6. trench

170.

| A | T | L | A | N | T | I | C | | | | |
|---|---|---|---|---|---|---|---|---|---|---|---|
| R | | | | | | N | | | | | |
| C | | | | | | D | | | | | |
| T | | | | | | I | | | | | |
| I | | | | | P | A | C | I | F | I | C |
| C | | | | | | N | | | | | |

171. Coriolis effect
172. bioluminescence
173. 4, 2, 6, 1, 5, 3
174. 1. the one without grass
    2. The grass held the soil in place and blocked some of the running water.
    3. Vegetation helps reduce soil erosion, especially on a slope or hillside.
    4. She could make sure she doesn't leave the soil bare, maybe by putting grass clippings on it.

175. 1. smog
    2. hydrocarbons
    3. photochemical smog
    4. acid rain

    Law: Clean Air Act

176. astronaut

| S | A | N | U | O | T | R | A | T |
|---|---|---|---|---|---|---|---|---|
| T | U | O | T | R | A | A | S | N |
| R | A | T | A | S | N | T | U | O |
| O | T | U | S | T | A | A | N | R |
| T | N | A | A | T | R | U | O | S |
| A | R | S | N | U | O | T | T | A |
| A | S | T | O | A | T | N | R | U |
| U | T | A | R | N | S | O | A | T |
| N | O | R | T | A | U | S | T | A |

177. Mercury, Venus, Earth, Mars, Jupiter, Saturn, Uranus, Neptune
178. 1. What is a galaxy?
     2. What is a constellation?
     3. What is a supernova?
     4. What is apparent magnitude?
179. Drawings should show Earth between the sun and moon so that the moon is in Earth's shadow.
180. 1. reflecting
     2. revolution
     3. elliptical
     4. solar flare